Transitions to Agriculture in Prehistory

Contributors

Ofer Bar-Yosef
Department of Anthropology
Harvard University
Cambridge, MA 02138

Anna Belfer-Cohen
Department of Archaeology
Hebrew University
Jerusalem
Israel

T. Michael Blake
Department of Anthropology and Sociology
6303 NW Marine Dr.
University of British Columbia
Vancouver, BC V6T 1Z1
Canada

Brian F. Byrd
Department of Anthropology
University of Wisconsin
Madison, WI 53706

Brian S. Chisholm
Department of Anthropology and Sociology
6303 NW Marine Dr.
University of British Columbia
Vancouver, BC V6T 1Z1
Canada

John Clark
Foundation for New World Archaeology
Brigham Young University
Salt Lake City, UT 84602

Angela Close
Department of Anthropology
Southern Methodist University
Dallas, TX 75275

Gary Crawford
Department of Anthropology
University of Toronto
Toronto ON M5S 1A1
Canada

R.E. Donahue
Department of Archaeology and Prehistory
University of Sheffield
Sheffield, England S10 2TN

Anne Birgitte Gebauer
Department of Anthropology
University of Wisconsin
Madison, WI 53706

Brian Hayden
Department of Archaeology
Simon Fraser University
Burnaby, BC V5A1S6
Canada

Larry Keeley
Department of Anthropology
University of Illinois
Chicago, IL 60680

Karen Mudar
Museum of Anthropology
University Museums Building
University of Michigan
Ann Arbor, MI 48109

T. Douglas Price
Department of Anthropology
University of Wisconsin
Madison, WI 53706

Fred Wendorf
Department of Anthropology
Southern Methodist University
Dallas, TX 75275

W.H. Wills
Department of Anthropology
University of New Mexico
Albuquerque, NM 87131

Transitions to Agriculture in Prehistory

Edited by
Anne Birgitte Gebauer
and T. Douglas Price

Monographs in World Archaeology No. 4

 PREHISTORY PRESS
Madison Wisconsin Kbs

Prehistory Press
7530 Westward Way
Madison, Wisconsin 53717-2009

James A. Knight, Publisher
Carol J. Bracewell, Managing Editor

ISBN 0-9629110-3-8
ISSN 1055-2316

Library of Congress Cataloging-in-Publication Data

Transitions to agriculture in prehistory / edited by Anne Birgitte Gebauer and
    T. Douglas Price.
            p.   cm. - - (Monographs in world archaeology, ISSN 1055-2316 ; no. 4)
    Papers presented at a symposium held at the 1991 meeting of the Society for
American Anthropology.
    Includes bibliographical references.
    ISBN 0-9629110-3-8 : $26.50
    1. Agriculture--Origin--Congresses. 2. Agriculture, Prehistoric--Congresses. I. Gebauer,
Anne Birgitte, 1951-   . II. Price, T. Douglas (Theron Douglas) III. Society for American
Anthropology. IV. Series.
GN799.A4T73 1992
306.3'49--dc20
                                                                                                    91-38392
                                                                                                    CIP

Cover art by Jørgen Mürhman-Lund.

8-6-93

Contents

Preface

The origins and spread of agriculture remain one of the more difficult problems in archaeology in spite of the great deal of work that has been done for many years, involving the development of both information and ideas.

The information we have on this process continues to increase. We now know for most areas what domestic plants and animals were important and when they first appeared—the time and space parameters of early food production. While a few places remain where this information is less readily available, basic information about the appearance and spread of domesticates is good. Ideas and explanations about why human groups adopted food production shortly after the end of the Holocene also continue to grow. Major theories today range from those involving the relationship between populations and their environments to social differentiation and competition.

This volume is an attempt to address new information on prehistoric transitions to agriculture in the context of explanations about why they took place. The papers in this volume were originally presented at a symposium on The Transition to Agriculture held at the 1991 meeting of the Society for American Anthropology. Because of the strong interest in the subject, as witnessed by the large attendance at the sessions, and because of the generally high quality of the presentations, we are making the papers available to a wider audience through this publication.

We would foremost very much like to thank the individual authors whose original work and ideas have made this volume possible. We would also acknowledge the talents of James Knight, Carol Bracewell, and Prehistory Press in bringing this volume to completion.

Anne Birgitte Gebauer
T. Douglas Price
Madison, Wisconsin
July 1991

Editor's note: The dates presented in this volume are uncalibrated unless otherwise noted by the authors.

1

Foragers to Farmers: An Introduction

Anne Birgitte Gebauer
T. Douglas Price
University of Wisconsin-Madison

Perhaps the most remarkable happening in our prehistory as a species was the almost simultaneous appearance and spread of domesticated plants and animals in many different areas of the world between about 10,000 and 5000 years ago. What is astonishing is the fact that this process of domesticating plants and animals appears to have taken place separately and independently in a number of different areas at about the same point in time. Given the long prehistory of our species, why should the transition to agriculture happen within such a brief period, within a 5000 year segment of the span of human existence?

The vast majority of our past was spent as hunter-gatherers. Our ancestry as food collectors, consuming the wild products of the earth, extends back at least four million years. Nevertheless, shortly after the end of the Pleistocene, some human groups began to produce food rather than collect it, to domesticate and control wild plants and animals. But the transition to farming involves much more than simple herding and cultivation. It also entails major, long-term changes in the structure and organization of the societies that adopt this new way of life, as well as a totally new relationship with the environment. Such a dramatic shift in the trajectory of cultural evolution demands understanding.

Explanations for the Origins of Agriculture

The origins and spread of agriculture remain one of the more difficult questions in archaeology in spite of the large number of studies that have been done since the end of the last century (e.g., Roth 1887). The possible causes of the transition to agriculture have been the subject of long and fervent discussion. The list of suggested causes for the "Neolithic Revolution" is almost endless (Table 1), but several major theories have dominated the debates. These major ideas can be described in terms of the important factors they emphasize: the oasis hypothesis, the natural habitat hypothesis, population pressure, human-plant symbiosis, or demands for social prestige. These ideas are really hypotheses about why domestication happened and can best be understood from an historical perspective, from early to more recent. The following discussion will briefly summarize some of the ideas that have had a major impact.

During the first half of this century, farming was thought to have originated on the dry plains of Mesopotamia where the early civilizations of the Sumerians and others arose. The best evidence for early farming villages at that time came from riverine areas or oases with springs in the Near East and Egypt, such as at Jericho in the Jordan Valley or along the River Nile in Egypt. At that time, the end of the Pleistocene was thought to be a period of increasing warmth and dryness in the earth's climate. Since the Ice Ages were cold and wet, the reasoning went, they should have ended with higher temperatures and less precipitation. Given this view of past climate, logic suggested that areas such as the Near East—a dry region to begin with—would have witnessed a period of aridity at the end of the Pleistocene when vegetation

Table 1. Some Suggested Causes for the
Transition to Agriculture.

aliens
big men
broad spectrum adaptation
circumscription
climatic change
competition
desertification
diffusion
domesticability
energetics
familiarity
fat intake
geniuses
hormones
intelligence
kitchen gardening
land ownership
multicausal
marginal environments
natural selection
natural habitat
nutritional stress
oases
plant migration
population growth
population pressure
random genetic kicks
resource concentration
resource pressure
rich environments
rituals
scheduling conflicts
sedentism
storage
technological innovation
water access
xenophobia
zoological diversity

grew only around limited water sources. The *Oasis* hypothesis then suggested a circumstance in which plants, animals, and man would have clustered in confined areas near water. Proponents of this idea, such as V. Gordon Childe (e.g., 1956), argued that the only successful solution to the competition for food in these situations would be for humans to domesticate and control both the animals and the plants. In this sense, domestication emerged as a symbiotic relationship for survival.

During the 1940s and 1950s, however, new evidence suggested that there had been no major climatic changes in the Near East at the close of the Pleistocene —no crisis during which various species would concentrate at oases. The new information forced a reconsideration of the origins of agriculture. Robert Braidwood suggested that domestication did not occur first in the lowlands of Mesopotamia. Braidwood (1960) pointed out that the earliest domesticates should appear in the *natural habitat* of their wild ancestors and that this area, the "hilly flanks" of the Fertile Crescent in the Near East, should be the focus of investigations. Braidwood and a large team of researchers excavated in northern Iraq. The evidence from the early farming village of Jarmo supported his hypothesis that domestication did indeed begin in the Natural Habitat. Braidwood did not offer a specific reason as to why domestication occurred, other than to point out that technology and culture were "ready" by the end of the Pleistocene, that humans were "familiar" with the species that were to be domesticated. At that time, farming was considered to be a highly desirable and welcome invention that provided security and leisure time. Archaeologists thought that once human societies recognized the possibilities of domestication, they would immediately start farming.

Lewis Binford challenged these ideas in the 1960s. Binford (1968, 1984) argued that farming in fact was back-breaking, time consuming, and labor intensive. He pointed to studies of living hunter-gatherers that indicated that these groups spent only a few hours a day obtaining food; the rest of the time was for visiting, talking, gambling, and the general pleasures of life. Such groups have been called "the original affluent societies." Even in very marginal areas today, such as the Kalahari Desert of southern Africa, food collecting is a successful adaptation and people rarely starve. Binford thus argued that human groups would not become farmers unless they had no other choice—that the origin of agriculture was not a fortuitous discovery, but a last resort.

Binford made his point in terms of an equilibrium between people and food, a balance that could be upset either by a decline in available food or an increase in the number of people. Since climatic and environmental changes appeared to be minimal in the Near East, Binford thought that it must have been an increase in the number of people that upset the balance. *Population pressure* was thus introduced as a causal agent for the origins of agriculture - more people, more mouths to feed, requiring more food. The best solution to that problem lies in domestication, which provides higher yields of food per acre of land. At the same time, of course, agricultural intensification also requires more labor to extract the food.

Kent Flannery (e.g., 1973) revised Binford's population argument slightly with the suggestion that the effects of population pressure would be felt most strongly not in the core of the natural habitat zone, where dense stands of wild wheat and large herds of wild sheep and goats were available, but rather at the margins of this zone where wild foods were less abundant. These ideas incorporating population pressure and the margins of the Fertile Crescent have become known as the *Marginal Zone* or *Edge* hypothesis.

Binford's concern with population was elaborated by Mark Cohen in a book entitled *The Food Crisis in Prehistory*, published in 1977. Cohen argued for an inherent tendency for growth in human population, a pattern responsible for the initial spread of the human species out of Africa, the colonization of Asia and Europe, and eventually of the New World as well. After about 15,000 B.C. or so, according to Cohen, all of the inhabitable areas of the planet were occupied and population continued to grow. At that time there was an increase in the use of less preferred resources in many areas. In addition to large game, land snails, shellfish, birds, and many more plant species were incorporated into the diet around the end of the Pleistocene, after 10,000 years ago. Cohen argues that the only way for a very successful, but rapidly increasing, species to cope with declining resources was for them to begin to cultivate the land and domesticate its inhabitants, rather than simply to collect the wild produce. Agriculture for Cohen then is a solution to problems of overpopulation on a global scale.

More recently others have argued that we cannot understand the transition to farming and food storage and surplus simply in terms of environment and population. These individuals have developed a *Social* theory for explaining the origins of agriculture. Barbara Bender (1978, 1990), for example, has suggested that the success of food production may lie more in the ability of certain individuals to accumulate food surplus and to transform those foods into more valued items such as rare stones and metals. From this perspective agriculture was a solution to a social problem. Roy Brunton may have provided some of the impetus for this perspective in his 1975 article titled "Why the Trobriands have chiefs?". The Trobriand Islanders have neither exceptional population density nor agricultural productivity. Brunton argues it is a participation in a closed system of kula exchange that limits the range of people who can effectively compete for leadership. Such a situation results in the emergence of a few big men who encourage the creation of surplus. Brian Hayden in a recent paper (1990) has argued specifically that it is the competitive and feasting aspects of Big Man rivalry that are the driving force behind food production. Nevertheless, the suggestion of social inequality, competitive feasting, and the emergence of big men as causes for the adoption of agriculture seems still to be a chicken/egg

argument. We simply do not yet have sufficient evidence to see whether agriculture precipitated such competition or vice versa, or whether the process was simultaneous.

These major theories represent an evolution in our thinking about and in our understanding of the transition. In a sense the theory building process is one of defining some of the more important variables involved in the emergence of farming societies. These variables obviously include factors such as the environment, climatic change, a broad spectrum resource base, population size and density, circumscription, resource availability and stress, social differentiation, the types of plants and animals present, and others.

But population, climatic change, or circumscription are impossible to indict as immediate and direct causes of change. More people *per se* do not dictate the adoption of agriculture; how are more mouths to feed directly translated into the cropping of hard-grained cereals? What advantages did the domestication of plants and animals offer to societies or individuals that brought about its adoption? Certainly factors such as population and climatic change may play a role in cultural evolution. But we cannot yet say precisely why plants and animals began to be domesticated shortly after the end of the Pleistocene. Some theories may seem reasonable in one of the primary centers of domestication, but not in another.

Simply put, there is no single, accepted, general theory for the origins of agriculture. The how and why of the Neolithic transition remains one of the more intriguing questions in human prehistory. The hypotheses that have emerged to date have helped to define some of the important variables in the transition but none seem to help us understand exactly why foragers turned to farming.

The Papers in This Volume

This volume, then, is an attempt to examine recent archaeological data on the transition to agriculture in light of existing theories. The subsequent chapters of this volume examine the transition in more detail in several different areas. Much of the discussion concentrates on the Near East and Europe because of the quantity and quality of archaeological information from this area. It is our intent to present the work and ideas of primary field workers—individuals familiar with the limitations of archaeological data and aware of the variability present in the archaeological record. The following paragraphs briefly describe the contents of the volume and the foci of the individual papers. The last segment of this introduction will summarize some of the findings of the authors.

Brian Hayden, *Contrasting Expectations in Theories of Domestication*

Hayden evaluates two major, competing hypotheses on the origin of agriculture: the "population pressure" model and the "competitive feasting" model. A number of contrasting expectations can be derived from the two models. The "competitive feasting" model of Hayden suggests that generalized hunter-gatherers maintain population levels in dynamic equilibrium with available resources. The reliance on unpredictable resources vulnerable to over-exploitation makes it essential for these societies to share food, and to have equal access to resources. The obligatory sharing of food renders increased labor investments in food production meaningless. Mobility rather than increased labor is the solution to inadequate resources.

Food production would therefore only occur where fundamental resources had become abundant, reliable, and invulnerable to over-exploitation. Under these conditions obligatory sharing would no longer be essential for survival and ownership of food no longer a taboo. This development would largely be the result of the development of mesolithic technology and the use of species invulnerable to over-exploitation. Hunter-gatherer societies with a highly developed technology living in areas of abundant resources are usually characterized by dense populations, semi-sedentism, socio-economic inequalities, labor-intensive status display and regional exchange of exotics. Thus domestication should appear first in such complex hunter-gatherer societies.

Hayden suggests that food production may have developed in the context of competitive feasting among ambitious individuals in the richest resource areas. The labor intensive production of domesticated species could be explained if the first domesticated species were used as delicacies or exotic foods at competitive feasts. In this scenario the domesticated foods would demonstrate the leader's success and power to control the labor of other people.

The expectations related to the competitive feasting model suggest that domestication would occur first in areas of rich resources. No population pressure is evident prior to domestication. Domesticated foods should be items suited for feasting rather than staple items. Also, domesticated foods would not expand beyond feasting realm before their productivity were competitive with other food sources. Domestication should occur first among complex hunter-gatherers, with food production most readily adopted by societies with competitive feasting and socio-economic hierarchy. Hayden finds that archaeological and ethnographic data from around the world support the feasting model, rather than population pressure.

Ofer Bar-Yosef and Anna Belfer-Cohen, *From Foraging to Farming in the Mediterranean Levant*

Bar-Yosef and Belfer-Cohen review the recent evidence on the transition to farming in one of the primary centers for the origins of agriculture, the Near East. This area contains the earliest evidence for agriculture and the largest number of domesticated plant and animal species. The review begins with the Geometric Kebaran, ca. 14,500 years ago, and goes through the earliest part of the Neolithic, the Pre-Pottery Neolithic A, which dates from ca. 13,000 to 10,500 B.P.

Paleoenvironmental data on climate, flora, and fauna provide the background to domestication in the Levant. Given a pronounced series of fluctuations in temperature and especially precipitation at the end of the Pleistocene, the transition to agriculture must be understood at least in part in terms of a response to changing environments. However, the Mediterranean zone of the Levant remained relatively stable and the impact of these fluctuations was minimal. This is the area where wild wheats and barley grow in abundance and where the earliest farming communities appeared 10,300 years ago.

The background to the emergence of these communities is found in the Geometric Kebaran, the Mushabian, the Natufian, and Khimian complexes. The Geometric Kebaran is found throughout the area of the Levant and dates from 14,500 B.P. to 13,000 B.P.; the Mushabian is found more to the south and continues somewhat longer in time to perhaps 11.000 B.P. Sites from this time are generally small and likely reflect small and mobile groups of hunter-gatherers utilizing the wild plants and animals of the region.

The major change in human adaptation in this area takes place shortly after 13,000 B.P. with the appearance of the Natufian complex, with a homeland in the central Levant—northern Israel and Jordan and southwestern Syria. The early part of the Natufian witnessed a shift to larger, permanent settlement, the extensive exploitation of wild cereals, the use of storage pits for food, the use of sickles, and a variety of groundstone tools. Outside the central Levant, a seasonally mobile pattern of hunting and gathering continues. The first fully Neolithic villages appear in Mediterranean Levant and are characterized by even larger communities and the use of domesticated plants. Agriculture appears when intentionally planned cultivation begins on a significant scale.

Bar-Yosef and Belfer-Cohen view the transition to agriculture in the Near East as very rapid, but one which began with the shift to more permanent settlement and the intensive utilization of wild stands of wheat and barley in the Natufian. The authors propose a complex model for the beginnings of agriculture which involves the pronounced climatic fluctuations of the end of the Pleistocene, the availability of cereals and legumes in the Mediterranean Levant, the development of technology of plant food collection and processing, the behavior of game animals, and the increasing numbers of people in the Levant at a time when every ecozone was already inhabited.

Brian F. Byrd, *The Dispersal of Food Production Across the Levant*

Byrd's paper takes off from the preceding survey by Belfer-Cohen and Bar-Yosef. Byrd considers the spread of agriculture *within* the Levant. This is a most interesting topic and one that has important implications for the spread of agriculture everywhere. Cytogenetic studies suggest that domestication of a number of founder crops (einkorn wheat, emmer wheat, barley, lentil, broad bean, chick pea, field pea, bitter vetch and flax) took place as single events. Each species was either domesticated in a different place or several species were domesticated together. The south-central Levant, where evidence of crop cultivation has been found at PPNA villages as early as 10,000 b.p., is the most probable center for these domestication events. The spread from this area is the subject of Byrd's paper.

Byrd points out that the expansion of food production into different ecological niches within Southeast Asia happened at different rates and with varying degrees of pervasiveness into local economies. Outside of a few PPNA villages in the south-central Levant, domesticated plants were integrated in what were essentially hunter-gatherer economies and appear to have been used only as a supplement in the diet for a prolonged period. Reliance on domesticated plants increased steadily through PPNB in the fertile area of the Levant and on the Jordanian and Syrian plateau. In other areas of Southwest Asia at this time, domesticated plants appear to play only moderate role or were absent.

Domesticated plants appear to precede the domestication of ovicaprids everywhere in the Levant by several hundred years. Herding of ovi-caprids was initially integrated into the local economy at sites along the forest-steppe boundary of the Central Levant between 9000–8000 b.p., but animal husbandry did not appear until later in Western Levant and its arid margins.

Sedentary village life emerged in most areas toward the end of PPNB when medium and large villages are found throughout the fertile area of the Levant and on the Jordanian and Syrian Plateau. A shift from round to rectangular architecture may document this change in settlement mobility. The considerable time lag between the onset of food production and the appearance of sedentary villages suggests that factors other than food production were important in the emergence of sedentism.

Interaction and exchange of information and possibly materials appear to be important mechanisms in dispersal of domesticates. Byrd hypothesizes that the initial spread of domesticates during PPNA took place through local adoption by indigenous hunter-gatherers in the fertile areas. Later during PPNB, both colonization by farming communities and local adoption by indigenous hunter-gatherers were responsible for the spread of food production.

Angela E. Close and Fred Wendorf, *The Beginnings of Food Production in the Eastern Sahara*

Close and Wendorf argue for the earliest independent domestication of cattle in the Eastern Sahara as a response to an ecological crisis and as a means for populations to expand into a new resource area. The ecological crisis was spawned by changes in the flow of the river Nile and the consequential narrowing of the floodplain. These environmental changes greatly reduced the availability of the most important late Paleolithic plant foods and catfish. Rainfall in the southeastern Sahara during the early Holocene opened up a new region into which the occupants of the Nile Valley could expand.

This desert adaptation is an example of local symbiosis between people and cattle. Herders could force the cattle to move often and far enough, and in the right direction, to find sufficient grazing and open water. Cattle provided the people with a reliable source of protein in the form of milk and blood. Together the two species survived in a very marginal environment, where likely both were unable to survive alone.

The suggestion that the cattle found at the desert sites were domesticated is based upon an ecological argument. Bone assemblages at the desert sites include non-drinking animals like hare and small gazelle, associated with cattle, but no intermediate size forms. This combination of species is highly improbable in ecological terms since cattle need to drink every day, or at least every second day, to survive. The limited amount of rainfall meant that no permanent standing water would have been available in the southeastern Sahara during the Holocene. During most seasons water could only be obtained by digging wells. Thus cattle could not have survived outside the Nile Valley without human support. Nor could the humans survive in that area without the supply of protein from the cattle.

Randolph E. Donahue, *Desperately Seeking Ceres:*
A Critical Examination of Current Models for the
Transition to Agriculture in Mediterranean Europe

Donahue reviews the current perspectives on the introduction of agriculture from the Near East into the Mediteranean basin. Conventional wisdom indicates that groups of colonists from southwest Asia along with their domesticated plants and animals, pottery, and village architecture diffused to the shores of the Mediterranean and southeastern Europe , bringing the Neolithic to Europe. The Ammerman/Cavalli-Sforza model for the advance of Neolithic colonists across Europe is one example of such thinking; Colin Renfrew's hypothesis for the spread of Indo-European languages by the first incoming farmers is another.

While it is clear that the majority of domesticated species must have come originally from the Near East, Donahue argues that there is no longer reason to assume that they were brought to Europe by foreign colonists. Rather he points to the likelihood of indigenous Mesolithic groups playing an active role in adopting the technologies, species, and materials associated with the Neolithic. Domesticated sheep and ceramics appear in the western Mediterranean by 7800 b.p., almost 1000 years prior to the cultivation of plants and other domesticated animals and technologies recognized as fully Neolithic. Donahue argues that sheep-herding was adopted as a risk-reducing strategy, and did not involve major changes in society. Small-scale pastoralism was integrated in a pattern of transhumance involving upland and lowland settlements by the mesolithic hunter-gatherers.

Lawrence H. Keeley, *The Introduction of Agriculture*
to the North European Plain

Keeley deals with an unusual situation in the spread of agriculture, a clear case of colonization by farmers of a new area. In most instances, agriculture seems to have spread by diffusion of products and ideas rather than the movement of people. Central Europe, however, witnessed a remarkably rapid spread of a group known from their pottery as the Linearbandkeramik. These farmers spread hundreds of kilometers to the east, north, and west from a center in Hungary within a period of less than 100 years. Keeley's concern is thus with the interaction between these invading farmers and the indigenous hunter-gatherers with whom they came in contact. In contrast to an earlier view which regarded this spread as peaceful, Keeley documents the hostility and absence of interaction between these two diverse groups of foragers and farmers. In addition to evidence for warfare, seen in the fortification of farming hamlets,

the indigenous foragers obtain domesticated animals and pottery from elsewhere, rather than the adjacent Linearbandkeramik communities. Keeley further documents the economic specializations that integrate and bind farming communities together in this rather hostile setting.

T. Douglas Price and Anne Birgitte Gebauer,
The Final Frontier: Foragers to Farmers
in Southern Scandinavia

Price and Gebauer consider the evidence for the transition to agriculture in northern Europe as one of the last areas in Europe to become "Neolithic". They begin with a summary of current information on the late Mesolithic and early Neolithic and a description of four stages in the transition from foragers to farmers: pristine hunter-gatherers, first contact with farmers, the adoption of domesticates and Neolithic pottery, and full scale farming communities. The information from northern Europe, while very detailed, remains puzzling because the nature of the transition to agriculture is two fold. On the one hand it is very gradual as seen in these four stages; on the other hand, the appearance of domesticates and Neolithic pottery seems to occur throughout southern Scandinavia almost simultaneously shortly before 3000 b.c.

Following this presentation of the evidence, Price and Gebauer consider current theories on the transition to agriculture in this area. These theories invoke sources for the "Neolithic" ranging from colonization by incoming Neolithic groups to the indigenous adoption of domesticates and the other accoutrements of farming communities. Proposed causes for the transition include environmental change, population pressure and social differentiation. The essay concludes with a discussion of some of the important variables and their role in the transition. These variables include climate, change in sea level, environmental change, population size and density, circumscription, resource availability and food stress, and social differentiation. While it is impossible to completely rule out any of these factors, population pressure and resource availability may not be as important as previously thought.

Gary W. Crawford, *The Transitions to Agriculture*
in Japan

This paper reviews the rather lengthy introduction and spread of agriculture across the Japanese archipelago from approximately 4000 B.C. until A.D. 1000—the Jōmon through the Yayoi periods and the replacement of the last Jōmon cultures by the ancestors of the Ainu—in terms of four stages or transitions.

Crawford documents the current evidence for plant utilization in the early Jōmon, including knotweeds, grasses, and sumac. Indigenous barnyard grass/millet appears to have been domesticated and cultivated in gardens by 2000 B.C. during the Middle Jōmon. In a second stage, wet rice was introduced from mainland Asia in the Yayoi period, around 400 B.C. Wet rice was common only in southern Japan; it was not until the first millennium A.D. that a distinctive pattern of agriculture involving millets, wheat, barley, and other dry field crops emerged in northern Japan. Finally, by 1000 A.D., intensive food production reaches Hokkaido following the breakdown of the last Jomon cultures.

The picture of agricultural transitions in prehistoric Japan is complex, with both indigenous domesticates and introduced cultigens, immigration from the mainland and acculturation of indigenous populations. The transitions are also mitigated by environmental and cultural factors: wet rice cultivation to the south and a mixed cereal economy to the north. Such studies document the complex nature of the origin and spread of agriculture and argue for detailed local investigations to enhance our understanding of the actual processes involved.

Michael Blake, John E. Clark, Brian Chisholm, and Karen Mudar, *Non-Agricultural Staples and Agricultural Supplements: Early Formative Subsistence In The Soconusco Region, Mexico*

Blake, Clark, Chisholm, and Mudar consider the transition to agriculture in the Formative period of coastal Mesoamerica (from approximately 1500 B.C. to the birth of Christ), specifically along the Pacific coast of Chiapas, Mexico. They review the evidence from this area in terms of two competing hypotheses: the competitive feasting model of Hayden (1990) and the interaction of plants and humans as described by Rindos (1984) and Flannery (1986).

Corn and a number of other plants were originally domesticated in the highlands of Mesoamerica. The important questions in this study are when, how, and why did these cultigens come to the environmentally stable and resource-rich coastal areas. The Pacific coast is an area of high biotic diversity and low seasonal variability. Information from settlement pattern, architecture, ceramics, and the distribution of trade goods such as obsidian and rare stones document the emergence of permanent and possibly hereditary chiefdoms in this area after 1550 B.C. The authors examine a variety of data to determine the nature of subsistence in these Formative communities. Faunal remains emphasize the importance of the freshwater estuarine resources including a wide variety of fish and reptiles, along with some terrestrial mammals. Paleobotanical

evidence documents the presence of domestic plants by 1350 B.C. Stable carbon isotope analysis of human bone from burials in this area, however, indicates that maize was not a significant part of the diet.

One intriguing aspect of the findings by Blake, Clark, Chisholm, and Mudar is the fact that these Early Formative chiefdoms on the Pacific coast of Mesoamerica were not agriculturally-based communities. Although these groups certainly consumed some maize, along with beans and other cultivated and wild plants, the faunal and isotopic evidence indicates that the estuarine resources were more important. This pattern continued in some areas well into the Middle Formative, while in other parts of the coastal zone maize became much more important in the diet at that time. Such local variation in responses to the availability of agriculture is also one of their conclusions.

W. H. Wills, *Foraging Systems and Plant Cultivation during the Emergence of Agricultural Economies in the Prehistoric American Southwest*

Wills discusses the introduction of maize, squash, and beans in the Late Archaic of the American Southwest. Wills argues that the initial use of these cultigens was as a risk avoidance behavior to enhance existing foraging systems. Maize was adopted because of its potential for storage and delayed use as a resource to be used in situations of stress. The adoption of these domesticated plants is at least indirectly related to an increasingly advantageous environment. The use of domesticates did not stimulate the emergence of large scale agricultural economies nor did it produce changes in the socioeconomic spheres of society. Indicators of overall economic intensification point toward a role of plant husbandry in facilitating increasingly localized economic pursuits. Contrary to the common wisdom, it does not appear that sedentism in the prehistoric Southwest developed along with dependence on agricultural surplus.

Wills distinguishes two modes of production, communal versus household, in the Late Archaic. The "communal" mode of production is characterized by group mobility, general reciprocity, limited degree of surplus production, diminishing returns, and poor capacity for economic intensification. The "household" mode represents a more sedentary way of life with restricted sharing, some reliance on surplus production, and both the incentive and the capacity for economic intensification. The household mode of production becomes more advantageous under competitive conditions where higher population densities and resource scarcity makes it desirable to control the use of key resources. However, the household mode is risk prone since the increased income is achieved by accepting higher probabilities of loss. Only individuals whose income is well above the minimum level necessary for survival are likely to adopt innovations with a high degree of risk or a higher variance of income. Wills suggests that a predictable and productive resource base is a necessary condition for restricted sharing and consequent enhancement of the potential for the economic intensification of food production. In other words the intensification of food production appears in situations of "plenty" rather than of scarcity. Sedentary life and restricted sharing are linked to the control of particular locations and competition for key resources, rather than the adoption of agriculture per se.

Conclusions

The first section of this introductory chapter reviewed some of the major theories concerning the origins and spread of agriculture. The second part of this chapter summarized the papers that appear in the following pages. It is now time to relate these two and to discuss what is new in this collection of papers in terms of perspectives on the transition to agriculture. Several important points emerge from the papers in this volume.

(1) Agriculture generally spreads through the diffusion of ideas and products rather than people. In almost every case in this volume, farming is adopted by indigenous peoples rather than brought in by colonists. Foragers become farmers rather than being replaced by them. The exceptions to this pattern are seen in the Linearbandkeramik of Central Europe and the expansion of cattle pastoralists into the Eastern Sahara. Colonization must now be seen as the exception rather than the rule in the spread of agriculture.

(2) Agriculture first appears in areas with an abundance of resources—the land of plenty—rather than a scarcity. The papers in this volume generally suggest that the transition to agriculture took place in areas with sufficient foods for the existing population. This seems to be the case almost everywhere from the Levant to Northern Europe to Japan to the Southwestern U.S. Wills makes the point that people already in an environment of risk will seldom try new subsistence strategies that bear even more risk. New strategies are initiated in situations where the risk is affordable. It is only in the cases of colonization, e.g. the Linearbandkeramik and the Eastern Sahara, where colonizing groups of agriculturalists or herders move into uninhabited areas where resources are scarce. The dense Atlantic forest of Central Europe and the rather dry reaches of the Eastern Sahara were resource poor without land clearance for cultivation or the herding of cattle.

(3) Agriculture appears and spreads quickly in areas where hunter-gatherers already occupy all of the inhabitable ecozones. This pattern can be seen virtually on a global scale in the rapid spread of agriculture in the Levant (Bar-Yosef and Belfer-Cohen, Byrd), across Europe (Donahue, Price and Gebauer), Mesoamerica (Blake et al.), the American Southwest (Wills), and elsewhere. The areas where colonization is seen further emphasize the packing of population that was present. Groups such as the Cattle Pastoralists of the Eastern Sahara or the Linearbandkeramik farmers of central Europe expand very quickly into ecozones with little or no human population as an alternative to areas which already contain hunter-gatherers. In this context social circumscription appears to be an important element of human life in the Late Pleistocene and early Holocene.

(4) Agriculture appears initially among more sedentary and complex groups of hunter-gatherers. In the Near East, for example, sedentism clearly preceeds the beginning of domestication. Elsewhere sedentary or semi-sedentary communities of foragers are the first to begin to experiment with cultigens and domesticated animals. The evidence from Mesoamerica has often been raised as the major exception to this pattern. Certainly the evidence from the southern Pacific Coast of Mexico (Blake et al., this volume), indicates that sedentary groups were present in the absence of significant agriculture. We would suggest that, in contrast to the common wisdom, the first farmers in Mesoamerica were likely semi-sedentary groups in resource-rich areas. The evidence from this important region remains sketchy at best and much more research needs to be done on the question of domestication. Complex hunter-gatherers in Japan, northern Europe, North and South America, are the first to shift to agriculture at the end of the Pleistocene (Price and Brown 1985).

(5) There is a long period of availability of cultigens and/or domesticated animals prior to full adoption of agriculture (Zvelebil and Rowley-Conwy 1986). Evidence from Europe and many other areas suggests significant and lengthy contact between foragers and farmers before agriculture is adopted by the hunter-gatherers. Many of the papers in this volume suggest that domesticates initially are a supplement or addition to existing foodstuffs and that considerable time lag is present before evidence for intensification of food production appears. Even in the Levant, the primary hearth of domestication in the Old World, there is considerable delay in the spread of domesticates during the early Holocene. Such evidence supports the idea that agriculture was initially

adopted in areas where food resources were not scarce or marginal.

(6) The transition to agriculture appears to be accompanied by a shift from a communal to household level of organization (Byrd, this volume). Such changes in community organization are noted in several areas and appear to reflect new modes of economic organization as well as a shift from communal sharing to familial or individual accumulation (Wills, this volume). The transition from egalitarian to hierarchical society may be witnessed immediately with the adoption of agriculture.

(7) There is a great deal of variability in local responses to the introduction of agriculture. This variability can be seen again in many areas (e.g., the Levant, the Southwestern U.S., Japan) where local environmental conditions play a major role in which domesticates were adopted and when. Such local variation likely obfuscates our understanding of the transition to agriculture and makes explanation more difficult.

(8) The general phenomena of the transition to agriculture is the same regardless of whether we are dealing with questions of primary origins or secondary spread. The question remains why did foragers become farmers. Foragers must have known and understood the propagation of plants and animal species for millennia. Evidence from many areas such as Jōmon Japan (Crawford, this volume), the Eastern U.S. (Smith 1989), the Near East (Belfer-Cohen and Bar-Yosef, this volume), and Europe (Donahue, this volume), documents the fact that foragers were utilizing certain plant species for a long period prior to the beginnings of cultivation. The important question is not whether certain groups were the first to domesticate specific plants or animals, but rather why they started to manipulate these species more intensively.

These conclusions mean that the basic assumptions of the current models for the transition to agriculture are incorrect and in need of revision. To better understand the question of why groups of foraging populations began to cultivate plants and herd animals at the end of the Pleistocene, it may be useful to distinguish among the consequences from, the conditions for, and the causes of the transition.

Certainly it is possible to identify the consequences of the shift to farming, that is to say what happened as a result of the adoption of agriculture. These consequences include the invention and adoption of new technologies for agriculture, the widespread use of ceramics, forest clearance and major, humanly-induced changes in the landscape, the cultivation of hard-shelled cereals, animal husbandry in many areas,

more villages and more people, and ultimately an increased pace along the path to more complex social and political organization. In fact the hard grained, storable nature of most cereals is likely the reason for their selection among the first domesticates. Such stored foods permit the accumulation of surplus, as well as providing a supply of food in lean periods.

We can also identify some of the conditions necessary for the transition to agriculture on the criteria of ubiquity; that is to say, farming generally does not begin unless at least some of these general conditions are present. On the basis of recent evidence these conditions would seem to include: (1) Sufficient population—not much seems to happen anywhere in terms of subsistence intensification until evidence for a substantial human presence appears in the archaeological record; (2) Some level of social circumscription—people in areas where agriculture spreads can no longer vote with their feet by migrating or retreating to unoccupied areas for new sources of food. Evidence for violent conflict is often present in areas where agriculture is initially developed or adopted; (3) Abundant resources—in every area where the adoption of agriculture is well documented, there appears to be a wide variety of foods available to the inhabitants. Farming generally does not appear under conditions of nutritional stress; (4) Potential or available domesticates—it goes without saying that agriculture cannot proceed without domesticates. It is, nevertheless, quite intriguing to see how many different plants were under cultivation or at least manipulated by hunter-gatherers shortly after the end of the Pleistocene.

The papers in this volume provide substantial insight into the necessary conditions for the transition. This new evidence emphasizes the difficulty of population estimates, the importance of social complexity, and the ubiquity of indigenous adoption. In contrast to earlier views, which argued for the origins of domestication among growing populations in marginal areas under resource stress, more recent evidence and the large majority of the papers in this volume suggest that the transition to agriculture took place in areas with substantial populations and with stable and abundant resources. Rather than the result of external forces and stress, the adoption of domesticates may well have been an internally motivated process. The consequences and conditions described above in fact suggest that human populations were pulled into the adoption of farming rather than pushed. But what or who pulled, and how?

This introduction ends without resolution of the question of causality. Perhaps, as Flannery said in his visit to the Master (1986: 512), "Why did agriculture begin? I'm not sure what I'd give you as a cause. And if you asked me,'What law did you come up with?' I'm not sure what I'd say." This question of *why* humans adopted farming remains elusive and at the same time one of the more intriguing in prehistory. It is our hope that the papers in this volume will improve your understanding of the problem, provide new insights into the question, and suggest directions for further research.

References Cited

Bender, B.
  1978 Gatherer-hunter to Farmer: a Social Perspective. *World Archaeology* 10: 204-222.
Bender, B.
  1990 The Dynamics of Non-hierarchical Societies. In *The Evolution of Political Systems*, edited by S. Upham, pp. 62-86. Cambridge University Press, Cambridge.
Binford, L.R.
  1968 Post-Pleistocene Adaptations. In *New Perspectives in Archaeology*, edited by S.R. Binford and L.R. Binford, pp. 313-341. Aldine, Chicago.
Binford, L. R.
  1984 *In Pursuit of the Past: Decoding the Archaeological Record.* Thames and Hudson, New York.
Braidwood, R.J., et. al.
  1960 *Prehistoric Investigations in Iraqi Kurdistan.* University of Chicago Press, Chicago.
Brunton, R.
  1975 Why do the Trobriands have Chiefs? *Man* 10: 544-558.
Childe, V.G.
  1956 *Piecing Together the Past.* Routledge & Kegan Paul, London.
Cohen, M.
  1977 *The Food Crisis in Prehistory.* Yale University Press, New Haven.
Crawford, G.
  1987 Ainu Ancestors and Prehistoric Asian Agriculture. *Journal of Archaeological Science* 14: 201-213.
Flannery, K. V.
  1973 The Origins of Agriculture. *Annual Review of Anthropology* 2: 271-310.
Flannery, K. V. (editor)
  1986 *Guilá Naquitz: Archaic Foraging and Early Agriculture in Oaxaca, Mexico.* Academic Press, New York.
Harris, D.R., and G.C. Hillman (editors)
  1989 *Foraging and Farming. The Evolution of Plant Exploitation.* Unwin Hyman, London.
Hayden, B.
  1990 Nimrods, Piscators, Pluckers and Planters: The Emergence of Food Production. *Journal of Anthropological Research* 9: 31-69.
Price, T.D., and J.A. Brown.
  1985 *Prehistoric Hunter-Gatherers: The Emergence of Cultural Complexity.* Academic Press, New York.
Rindos, D.
  1984 *The Origins of Agriculture: an Evolutionary Perspective.* Academic Press, New York.
Roth, H.L.
  1887 On the Origin of Agriculture. *Journal of the Royal Anthropological Institute of Great Britain and Ireland* 16: 102-136.
Smith, B.D.
  1989 Origins of Agriculture in Eastern North America. *Science* 246: 1566-1571.
Vaquer, J., D. Geddes, M. Barbaza and J. Erroux.
  1986 Mesolithic Plant Exploitation at the Balma Abeurador (France). *Oxford Journal of Archaeology* 5(1).
Zvelebil, M. , and P. Rowley-Conwy.
  1986 Foragers and Farmers in Atlantic Europe. In *Hunters in Transition. Mesolithic Societies of Temperate Eurasia and Their Transition to Farming*, edited by M. Zvelebil, pp. 67-93. Cambridge University Press, Cambridge.

2

Models of Domestication

Brian Hayden
Simon Fraser University

I have recently argued for an alternative explanation of the transition to agriculture that does not rely on population pressure, climatic change, or other relatively popular explanatory factors (Hayden 1990). What I would like to do here is apply standard scientific procedures for evaluating competing theories—procedures often not explicitly used in archaeology. These procedures involve examination of the contrasting types of expectations that can be derived from the model that I presented and other competing models of domestication. Because of the large number and variety of theories concerning the origins of agriculture, it is impossible to evaluate all theories in this fashion here. I therefore will concentrate on comparing my competitive feasting model with a single widely endorsed model of domestication involving population pressure. While I focus on population pressure in this paper, other models can be examined in a similar fashion.

Before beginning this comparison, it is worth noting as does Bettinger (1980:233), that simple manipulation of the productivity of wild species is commonplace among hunter-gatherers throughout the world, either via transport, fire, irrigation, sowing, transplanting, or stocking. Such strategies may have been used by hunter-gatherers throughout history. I suggest that what requires explanation is not this type of elementary manipulation, but intentional behaviors *calculated to change* the physical properties of the wild species.

The Population Pressure Model

As argued by Cohen (1977) and a number of others, human populations exhibit an inherent and poorly controlled natural tendency to increase in number. According to this model, when human populations are low, as at the initial stages of population growth or colonization of new regions, they can afford to employ the most efficient food-getting strategies available. However, as populations grow and density increases, the most easily obtained foods are quickly consumed. Driven by nutritional inadequacies, the growing numbers of people must either move to less populated areas, or they must exploit other more labor-intensive sources of food in order to provide food for everyone. Eventually, however, all potential natural habitats are filled up and populations are forced to use still more labor-intensive techniques to survive. This is represented by Mesolithic/Archaic adaptations in which small-seeded grasses or other labor-intensive foods are used. With still further population growth, populations continue to experience resource-population imbalances and stresses. At this point, according to the advocates of population pressure, communities are forced into the most labor-intensive subsistence strategy of all—food production. Nor does the cycle stop at this point, but it goes on to greater intensification as described by Boserup (1968). This entire scenario is built on classic Malthusian causality.

There are a number of very clear expectations that can be derived from the population pressure model.

1. There should be an increase in population pressure in the period immediately preceding domestication.

2. Domestication should occur first in those areas subject to the most stress. In general, these are marginal areas, such as the semi-deserts proposed by Binford (1968).

3. Because marginal areas are expected to be the first and most strongly affected, the cultures involved should be those of relatively simple hunter-gatherers with relatively poor resources and limited consequent population densities.

4. The first foods to be domesticated should be items that are important as staples, or at least items that would significantly help avert starvation.

5. Because population increase inexorably continues, once food production began there should be continued and constant pressure to increase the overall contribution of domesticated foods in diets.

6. In situations of contact between hunter-gatherers and agriculturalists, the diffusion and acceptance of food producing technology ought to take place most readily among hunter-gatherers living in the most stressed environments, assuming food production is a viable option in those environments (i.e., excluding arctic and fully desert environments).

These central features and expectations of the population pressure model are summarized in Table 1.

The Competitive Feasting Model

In contrast to constant uncontrolled population growth, I have long argued that generalized hunter-gatherers employ a strategy of maintaining their population levels in a dynamic equilibrium with their resources (Hayden 1972, 1986). In the face of semi-cyclical fluctuations in resources of varying magnitudes, this equilibrium is neither so high that starvation is a frequent event, nor so low that popula-

tion size adversely affects mating, defense, or constantly requires excessive population controls.

The limited and fluctuating nature of the resource base for most generalized hunter-gatherers almost mandates the sharing of food within the group and the establishment of inter-group alliances to obtain help in times of food shortage. The critical importance of sharing and the vulnerability of at least some staple resources to over-exploitation render claims of private ownership, competition based on economic resources, and egotistical behavior an anathema among generalized hunter-gatherers. Under these conditions, it makes no sense to invest any significant extra time or labor in food production since it is primarily others who will profit due to the obligatory sharing of food.

The only conditions, then, under which one might expect food production to occur would be where the fundamental resource characteristics had changed in such a way that obligatory sharing was no longer essential for survival and where ownership of produced food was no longer anathema. I have argued that these changes only take place where resources are abundant, more reliable, and invulnerable to over-exploitation (Hayden 1981, 1990, in press). Moreover, once these basic changes occur (largely due to techno-logical innovations), competition involving subsistence resources no longer has detrimental conse-quences for the resource base. I see these developments as emerging largely with the advent of Mesolithic technology (Table 2)—especially with either the exploitation of *r*-selected species or the massive culling and storage of meat from large herds of migrating species that could not be overexploited due to their sheer numbers (e.g., reindeer, whales).

Given these conditions, it is reasonable to assume that the competitive feast with hierarchical socioeco-nomic rivalries could have emerged in the richest resource areas. I suggest that many ambitious individ-uals used the economically based competitive feast to gain control over people's labor, loyalties, and loans. In competitive feasts, control over labor is a primary goal and it is used as a symbol of success and power. If

Table 1: Expectations of the Population Pressure vs. Competitive Feasting Model		
Population Pressure Predictions:	Criteria	Competitive Feasting Predictions:
1. Yes	Population Pressure	No
2. Stressed	Resource Character	Rich
3. Simple	Cultural Level	Complex
4. Staples	Nature of Domesticate	Feasting Food
5. Increasing	Subsequent Reliance on Domesticates	Stable Use
6. Stressed Environments	Diffusion Context	Rich Environments
7. n/a	Evidence for Feasting	Yes

the first domesticated species were highly desirable foods, but labor intensive to produce or otherwise difficult to obtain, it is most likely that domestication developed within the context of competitive feasting.

The expectations that can be directly derived from the competitive feasting scenario are:

1. There is no reason to expect significant increases in overall population pressure immediately prior to domestication, although once socioeconomic inequality became firmly established, disenfranchised segments of communities might experience resource stress on a recurrent basis.

2. Domestication should first occur in areas of rich resources, especially areas with abundant *r*-selected resources such as grasses and fish.

3. Domestication should first occur among more complex hunter/gatherer cultures with indications of dense populations, semi-sedentism, socioeconomic inequalities, labor-intensive status display items, and significant regional exchange in status display exotica.

4. There should also be evidence for feasting/ritual structures and/or areas.

5. The first domesticated foods should be items suitable for feasting rather than staples, i.e., they should be delicacies or items that would be labor intensive to produce or obtain.

6. Because competitive feasting would constitute an episodic and relatively infrequent event, there would be no reason for domesticated foods to expand beyond the realm of feasting at least until the productivity and net rate of return of domesticated foods increased to the point where they would be competitive with other food procurement strategies. Thus, domesticated foods

could be expected to occupy a very minor position for a relatively long time in the overall subsistence composition of a community.

7. Finally, in situations of contact between food producers and hunter-gatherers, the competitive feasting model predicts that hunter-gatherers with competitive feasts and socioeconomic hierarchies would adopt food-producing technology much more readily than any other type of hunter-gatherers.

These are the core elements and expectations of the competitive feasting model.

Tests from the Real World

If we compare these contrasting expectations to archaeological observations we find the following:

(1) The population pressure model posits signs of stress prior to domestication, whereas the competitive feasting model does not. Despite a world-wide search for indicators of population pressure immediately prior to the first evidence for domestication, Cohen and Armelagos (1984:592-596) and others (Roosevelt 1984:574,577; Meiklejohn and Zvelebil 1991; Buikstra et al. 1986) found scant support for this expectation. These results clearly favor the competitive feasting hypothesis and fail to support the expectations of population pressure models. Moreover, with the repeated and extreme climactic changes that characterized the Pleistocene, bands and entire regions must have been repeatedly subjected to resource-population stresses. Encroaching glaciers, sea levels, deserts and unproductive forests should have created regional population pressures during the Lower and Middle

---

Table 2: Mesolithic Archaic Technological Innovations and Resources*

Basketry, including water-tight varieties (used for seed collection, parboiling, soups to extract all fats and proteins from bones and meat).
Fire-cracked rock (for boiling plants and steaming open shellfish and snails).
Grinding stones, mortars and pestles (for processing grass seeds, toxic nuts like acorns, pounding up small animal bones and cartilage, pounding up woody nuts for boiling, etc.)
Nets (for fishing, capturing small game like rabbits, capturing birds such as fowl).
Weirs (for capturing fish).
Fishhooks and leisters (for fishing).
Basket traps (for fish and small game).
Elaborate snares and traps (for small and large land game, fish, and birds).
Bow and arrow (for hunting).
Microliths (for sickles, knives, arrowheads, and spearheads).
Domesticated dogs (for hunting and transport).
Firing of forest and plains over wide areas (to create more and better-quality grazing for animals and thereby increase their abundance).
Sleds and canoes (to increase access to distant places where important resources were located).

---

* Some of these innovations may have initially occurred in the late Upper Paleolithic, but appear to have been fully developed only in the Mesolithic period.

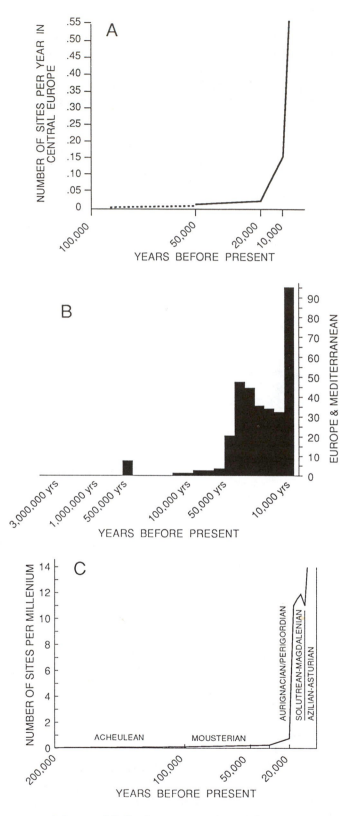

Figure 1. Graphs from three areas of the world displaying pronounced increases in site numbers and areas indicative of substantial increases of population. In the competitive feasting model, such population increases are viewed as responding to increases in extractable resources made possible by late Pleistocene and Holocene technological advances. **A** represents increases in the number of sites over time in Central Europe (after Vencl 1991); **B** represents increases in dated sites over time in Africa (after Isaac 1972); **C** represents increases in site numbers per millenium in Cantabrian Spain (after Clark and Straus 1983). Similar trends are documented from Paleo-Indian to Arcahaic times in the New World by Stark (1986).

Pleistocene similar to those at the end of the Pleistocene. Yet these stresses did not result in domestication during earlier periods of time. Resource-demography stresses seem rather to have characterized the entire Pleistocene at episodic intervals, but must have been resolved by means other than food production (e.g., fusion with other bands in conjunction with population curtailment).

It is also difficult to imagine why domestication should occur independently in the Old World and the New World within a few thousand years, since the Old World had witnessed growing population pressure for 2,000,000 years (with the African overflow severely constrained by the Suez isthmus); the New World was occupied for only about one hundredth of that time and immigration was severely constrained by the Bering Strait. In sum, the mechanics of population increase do not seem capable of accounting for the almost contemporaneous development of agriculture in these two areas.

(2) The population pressure model predicts that domestication will begin in marginal areas, whereas the competitive feasting model expects domestication to develop in rich resource areas. Some of the earliest evidence for domestication comes from areas that today are considered marginal, such as the Levant, Jōmon Japan, and the Valley of Oaxaca. It is clear, however, from the archaeological record that these regions were far from marginal at the end of the Pleistocene. While these areas may be difficult for rainfall farming today, at the end of the Paleolithic they supported rich stands of grasses, nuts, herds of gazelle or deer, and/or considerable aquatic life. On the basis of environmental reconstructions, semi-sedentary settlement patterns, faunal remains, and exponentially increasing densities of settlements in the periods leading up to domestication (Figure 1), it is clear that the exploited resource base of groups on the verge of domestication was quite abundant rather than marginal and highly stressed. Even on the seemingly impoverished slopes of the Valley of Oaxaca, Flannery (1986:314-5) calculates that over 3,000 tons of plant food and almost five tons of venison would have been available in the five km catchment area of a single camp, while Schoenwetter and Smith (1986:217) emphasize the "great quantities" of resources available in the valley bottoms in Oaxaca. Thus, these results largely follow the expectations of the competitive feasting model and fail to support the expectations of the population pressure hypothesis.

(3) Similarly, the population pressure model predicts impoverished, stressed, egalitarian cultures to first develop domestication; the competitive feasting model expects domestication to develop among more complex cultures with socioeconomic inequalities. The archaeological cultures where domesticates first appear generally exhibit considerable signs of control over labor, socioeconomic inequality, and exchange in exotic materials. These factors are far more consistent with an economic surplus than with economic deficits bordering on chronic starvation. In the Natufian, the Jōmon, the Eastern American Archaic, and to a lesser degree in the Oaxacan and Tehuacan Valleys, there are stone, bone, and shell bracelets; carved stone bowls and containers made from other unusual materials such as turtle shell and clay; bone and shell jewelry or hair pieces; special burials; human sacrifices (slavery?); special pigments and attractive minerals including copper; paint palettes; sculptures in various media; stone scepters; and elaborated residential structures (see Hayden 1990). These indicate considerable surpluses and control of labor. These results largely confirm the hierarchical socioeconomic predictions of the competitive feasting model and fail to corroborate the egalitarian expectations of the population pressure model.

(4) Although the population pressure model involves no expectations concerning special feasting areas or structures, the competitive feasting model does. The limited nature of most site excavations from this time period (as well as the general lack of ethnographic information on the topic) restricts the ability to test this expectation thoroughly. However, good candidates for such structures and areas certainly exist for the Natufian at Asiab and Beidha, and for the Jōmon where open plazas in the larger sites have been interpreted as loci of intercommunity rituals (see Hayden 1990). Thus, this expectation is partially confirmed for the competitive feasting model.

(5) Population pressure implies that the first foods to be domesticated should be staples; the competitive feasting model expects the first domesticates to be feasting foods (or feasting/status paraphernalia). The first foods domesticated in North and Central America and in Japan consist of chili peppers, gourds, squash, avocados, hemp, colza, mint, and burdock. None of these early cultigens can be considered to have significant value for averting starvation. They all fit the role of condiments and delicacies much better. The one major exception to this pattern (wheat and barley in the Near East) can be reasonably subsumed under the feasting scenario. Dietler (1990) has argued that these early grains, as well as maize, were originally more important in brewing than in subsistence in prehistory. He has demonstrated the pivotal role that beer production plays in the mobilization of labor by ambitious

and wealthy individuals in traditional societies. In fact, he emphasizes that it is only wealthy individuals that can afford to produce the grain surpluses and underwrite the labor required to convert grain into beer on a large scale. Therefore it is only the wealthy that can engage in beer-based competitive feasts or that can mobilize work parties based on beer.

Dogs constitute an additional domesticated species that makes little sense in terms of the population pressure model, but is comfortably accommodated as a status and possibly feasting item within the competitive feasting model. Domesticated gourds similarly cannot be explained by appealing to famine-related causality. On the other hand, special gourds are widely produced and used in feasting contexts, so that they, too, fit in with the competitive feasting scenario. Thus, the archaeological evidence largely supports the expectations of the competitive feasting model rather than the expectations of the population pressure model.

(6) The population pressure model leads to expectations regarding the constantly increasing importance of domesticates in subsistence economies over time. The competitive feasting model anticipates a long period of stasis after the introduction of the first domesticates. Such periods of stasis have been noted by many prehistorians and paleoethnobotanists as curiosities in the Eastern American Archaic, Jōmon, and the Near East (Crawford 1992, this volume; Smith 1989). Recent work by Blake, Chisholm, Clark, and Mudar (this volume) has shown that early complex sedentary communities in coastal Chiapas were largely foraging based, and that they consumed insignificant amounts of maize for about 700 years after this cultigen was first introduced. This makes a great deal of sense in the terms of the competitive feasting model in which domesticated foods are only used in occasional feasts. This is especially the case if Flannery is correct in his observations that Formative maize could not have produced a yield considered worthwhile for subsistence purposes by contemporary agriculturalists, i.e., that it was a labor-intensive food (Flannery 1973:297-8). Flannery estimates that the earliest maize from the Tehuacan Valley produced only 60–80 kg per hectare or less, whereas modern subsistence farmers consider yields less than 200–250 kg per hectare as not worthwhile. Flannery also estimates that the higher yields were only achieved in the second millennium B.C. when agricultural village life was first established, i.e., over 4,000 years after the first appearance of domesticated plants.

In what appears to be a related phenomenon, the synchronic, ethnographic examination of the importance of food production in traditional societies shows a curiously "anomalous" bimodal distribution according to Hunn and Williams (1982: Figure 2). This distribution is inexplicable in terms of the population pressure model; however it makes a great deal of sense if one assumes that the societies relying on less than 5% of domesticated foods are ones that use these foods primarily in competitive feasting contexts, due to the extra labor necessary to produce them. As Hunn and Williams (1982:5) state:"If agriculture were just another resource type, there would be no reason to expect such a dearth of cases in the range of 5 to 35 percent dependence." While the societies that use domesticated foods for less than 5% of their diets should do so because of the high labor costs (and only for feasts), the societies that rely more than 50% on domesticated foods for their subsistence should be those that have found ways of making the net return on agricultural production more efficient than the net return on foraging, probably through genetic manipulation or improvements in other social or technological factors.

In all, expectations of the competitive feasting model concerning the slight overall subsistence role of domesticates during protracted initial periods are again upheld, while those of the population pressure model do not conform to empirical data.

(7) Finally, population pressure models predict more ready and rapid diffusion of food production technology to hunter-gatherers under resource stress, while the competitive feasting model expects food producing to diffuse most rapidly among complex hunter-gatherers with socioeconomic hierarchies and competitive feasting. In both scenarios minimally suitable physical environments for domestication are required, thereby excluding extremely dry and extremely cold areas. A cursory examination of the spread of food producing among hunter-gatherers that were contacted by Europeans indicates that generalized hunter-gatherers such as those of Australia, the Great Basin, and the Kalahari were the slowest and most reluctant groups to take up growing of even the highly productive foods of industrial societies, in spite of prodding by government and missionary organizations. In contrast, within a 15 year period from 1827 to 1840, the cultivation of European potatoes spread throughout the coastal native communities of the American Northwest Coast through the initiative of elite members (Suttles 1951). The main use of these potatoes was as a feasting food. As in other areas such as the Levant, the Mediterranean, and northern Europe (see chapters in this volume by Byrd, Keeley, Donahue, and Price and Gebauer), food production appears to have diffused to certain Mesolithic groups relatively easily and rapidly and even from one

Mesolithic group to another, much as the potato diffused on the Northwest Coast. I would argue that this was determined by the socioeconomic complexity of these groups as well as the suitability of local environments for food production.

In fact, Meiklejohn and Zvelebil (1991:137) suggest that limited plant husbandry and cultivation was already present in some European Mesolithic communities well before Neolithic influences arrived, and that once Neolithic communities became established, a wide range of status goods were exchanged between certain Mesolithic and Neolithic groups. I suspect that the reason the Mesolithic and Archaic only lasted for a few thousand years in many areas of the world was that technological improvements of this period increased exploitable resource availability to such a degree that conditions were ripe for the establishment of complex hunter-gatherers with economically based competition and competitive feasting systems. Once established, these conditions provided fertile ground

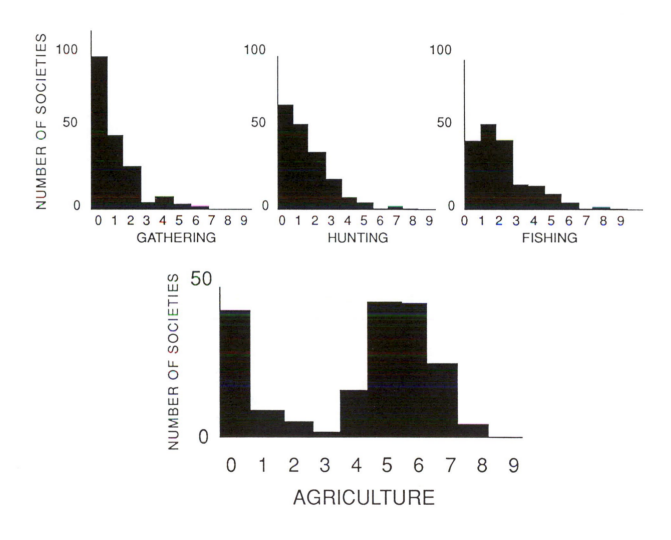

Figure 2. Graphs showing the relative dependence of traditional societies on gathered foods, hunting, fishing, and agriculture (from Hunn and Williams, 1982: 6). The aberrant, bimodal characteristic of reliance on agriculture is clearly evident here, indicating that societies with very low reliance on agriculture use domesticated foods in special contexts only, such as competitive feasting. The sample is drawn from Murdock's *Ethnographic Atlas*. Numbers on the ordinate indicate the number of societies out of 200 with a given reliance on a particular type of food procurement; numbers on the abscissa indicate the relative degree of reliance as defined by Murdock (1 = 0–5%; 2 = 6–15%.... 9 = 86–100%). Copyright 1982 by AAAS.

for the spread of food production. Variations in the initial spread of domestication, such as those documented by Bar-Yosef and Belfer-Cohen (this volume) can be explained by local or regional differences in underlying resource potentials for supporting socioeconomic complexity and competitive feasting (in conjunction with local variations in suitable climates for growing food).

Thus, impressions from contact ethnographies and from prehistoric patterns of diffusion tentatively appear to conform better to the expectations of the competitive feasting model than to the expectations of population pressure hypotheses. I would like to emphasize, however, that the competitive feasting model does not entirely preclude the diffusion of domestication to generalized hunter-gatherers. What the competitive feasting model does predict is that such diffusion should only occur when the net return of domesticated foods is equal to, or greater than, the net return on harvested wild foods. In general, as Flannery's data seem to dictate, early domesticates appear to have been highly labor intensive and could likely only compete with the returns of wild foods after hundreds or thousands of years of genetic manipulation.

Conclusions

The population pressure model and the competitive feasting model for the domestication process provide a number of striking contrasts in terms of their expectations of current archaeological evidence. It is possible to equivocate about some of these predictions and to develop auxilliary hypotheses to accomodate discrepancies between expectations and empirical observations. However, several of the differences in expectations are so pronounced that any attempt to accomodate them seems unrealistic.

Auxilliary scenarios might conceivably be developed to account for the lack of evidence for increases in population pressure, or to negate the necessity of initial domestication occurring in marginal habitats, although the expectations as I have stated them are those developed by population pressure proponents. However, it seems beyond reason to explain the condiment-delicacy-feasting nature of many first domesticates in terms of any strategy to allay significant food shortages. Nor is there any simple fashion in which the population pressure model can explain the bimodal distribution of reliance on food production in the ethnographic record (Figure 2).

The population pressure model of domestication has a number of critical flaws. In comparing expectations between it and the competitive feasting model, population pressure consistently fails to be supported by empirical data, while the competitive feasting model exhibits a relatively good fit. There are, of course, other models that attempt to account for domestication (e.g., Bender 1978, 1985a, 1985b). These, however, appear to compare even less favorably with either of the two models discussed here. Nevertheless, detailed comparisons of the relative theoretical qualities and expectations of these other models cannot but help advance our understanding of one of the most important transitions in human prehistory.

References Cited

Bender, Barbara
>1978 Gatherer-hunter to Farmer: A Social Perspective. *World Archaeology* 10:204-222.
>1985a Emergent Tribal Formations in the American Midcontinent. *American Antiquity* 50:52-62.
>1985b Prehistoric Developments in the American Midcontinent and in Britanny, Northwest France. In *Prehistoric Hunter-Gatherers*, edited by T.D. Price and J.A. Brown, pp. 21-57. Academic Press, Orlando, Florida.

Bettinger, Robert
>1980 Explanatory/Predictive Models of Hunter-gatherer Adaptation. *Advances in Archaeological Method and Theory* 3:189-255.

Binford, Lewis
>1968 Post-Pleistocene Adaptations. In *New Perspectives in Archaeology*, edited by S. Binford and L. Binford, pp. 313-342. Aldine, Chicago.

Boserup, Ester
>1965 *The Conditions of Agricultural Growth*. Aldine, Chicago.

Buikstra, Jane, Lyle Konigsberg, and Jill Bullington
>1986 Fertility and the Development of Agriculture in the Prehistoric Midwest. *American Antiquity* 51:528-546.

Clark, Geoffrey, and Lawrence Straus
>1983 Late Pleistocene Hunter-gatherer Adaptations in Cambrian Spain. In *Hunter-gatherer Economy in Prehistory*, edited by G. Bailey, pp. 131-148. Cambridge University Press, Cambridge.

Cohen, Mark
   1977 *The Food Crisis in Prehistory*. Yale University Press, New Haven.
Cohen, Mark, and George Armelagos
   1984 Paleopathology at the Origins of Agriculture: Editor's Summary. In *Paleopathology at the Origins of Agriculture*, edited by M. Cohen and G. Armelagos, pp. 585-602. Academic Press, Orlando, Florida.
Crawford, Gary
   1992 Plant Remains from Carlston Annis (1972, 1974), Bowles and Peter Cave. Kent State University Press, Kent, Ohio, in press.
Dietler, Michael
   1990 Driven by Drink: The Role of Drinking in the Political Economy and the Case of Early Iron Age France. *Journal of Anthropological Archaeology* 9:352-406.
Flannery, Kent
   1973 The Origins of Agriculture. *Annual Review of Anthropology* 2:271-310.
Flannery, Kent (editor)
   1986 *Guilá Naquitz: Archaic Foraging and Early Agriculture in Oaxaca, Mexico*. Academic Press, Orlando, Florida.
Hayden, Brian
   1972 Population Control Among Hunter-gatherers. *World Archaeology* 4:205-221.
   1981 Research and Development in the Stone Age: Technological Transitions Among Hunter-gatherers. *Current Anthropology* 22:519-548.
   1986 Resources, Rivalry and Reproduction: The Influence of Basic Resource Characteristics on Reproductive Behavior. In *Culture and Reproduction*, edited by Penn Handwerker, pp. 176-197. Westview Press, Boulder, Colorado.
   1990 Nimrods, Piscators, Pluckers and Planters: The Emergence of Food Production. *Journal of Anthropological Archaeology* 9:31-69.
   1992 Conclusions: Ecology and Complex Hunter-gatherers. In *Complex Cultre of the British Columbia Plateau*, edited by B. Hayden. University of British Columbia Press, Vancouver, in press.
Hunn, Eugene, and Nancy Williams
   1982 Introduction. In *Resource Managers: North American and Australian Hunter-gatherers*, edited by N. Williams and E. Hunn, pp. 1-16. Westview Press, Boulder, Colorado.
Isaac, Glynn
   1972 Chronology and The Tempo of Cultural Change During the Pleistocene. In *Calibration of Homonid Evolution*, edited by W. Bishop and J. Miller, pp. 381-430. University of Toronto Press, Toronto, Ontario.
Meiklejohn, Christopher, and Marek Zvelebil
   1991 Health Status of European Populations at the Agricultural Transition and The Implications for the Adaptation of Farming. In *Health in Past Societies*, edited by H. Bush and M. Zvelebil, pp. 129-145. BAR International Series. Oxford.
Roosevelt, Anna
   1984 Population, Health, and The Evolution of Subsistence: Conclusions from the Conference. In *Paleopathology at the Origins of Agriculture*, edited by M. Cohen and G. Armelagos, pp. 559-584. Academic Press, Orlando, Florida.
Schoenwetter, James, and Landon Smith
   1986 Pollen Analysis of the Oaxaca Archaic. In *Guilá Naquitz*, edited by K. Flannery, pp. 179-237. Academic Press, Orlando, Florida.
Smith, Bruce D.
   1989 Origins of Agriculture in Eastern North America. *Science* 246: 1566–1571.
Stark, B.
   1986 Origins of Food Production in the New World. In *American Archaeology, Past and Future*, edited by D. Meltzer, D. Fowler, and J. Sabloff, pp. 277-321. Smithsonian Institution Press, Washington, D. C.
Suttles, Wayne
   1951 The Early Diffusion of the Potato Among the Coast Salish. *Southwestern Journal of Anthropology* 7:272-288.
Vencl, Slavomil
   1991 On the Importance of Spatio-temporal Differences in the Intensity of Paleolithic and Mesolithic Settlement in Central Europe. *Antiquity* 65: 308-17.

3

From Foraging to Farming
in the Mediterranean Levant

O. Bar-Yosef
Peabody Museum, Harvard University

A. Belfer-Cohen
Institute of Archaeology, Hebrew University

The subject of this paper has been dealt with extensively in many recent publications (e.g., Moore 1985; Bar-Yosef and Belfer-Cohen 1989, 1991; Henry 1989; McCorriston and Hole 1991). However, the continuous accumulation of newly published data requires a novel synthesis every few years. This paper aims at filling in some of the gaps in previous papers while discussing a few major issues of the process which led to the emergence of agriculture as a new subsistence strategy. In the course of presenting the data we prefer to separate the archaeological observations from the interpretations. We also choose to begin the description of the archaeological entities with the Geometric Kebaran complex and end it with the earliest Neolithic, also known as the "Pre-Pottery Neolithic A" period in the southern Levant (Figure 1). The detailed archaeological sequence as well as the basis for the identification of the social units within the archaeological entities have been discussed in detail elsewhere (Bar-Yosef and Belfer-Cohen 1989).

The Region

The Mediterranean Levant is a small region in southwestern Asia: it is about 1100 km long and about 250–350 km wide. It stretches from the southern flanks to the Taurus Mountains in Turkey into the Sinai Peninsula, with its eastern border marked by the Middle Euphrates Valley, Palmyra Basin, Gebel ed-Druz, the Azraq, and El-Jafr Basins. The variable topography of the region, described elsewhere (Bar-Yosef and Belfer-Cohen 1989), includes the coastal ranges, the Rift Valley, the inland mountain ranges, and the eastward sloping plateau dissected by many wadis.

The climate of the Levant is dominated by two seasons: cold, rainy winters and hot, dry summers. On the largely limestone and chalky rocks, where annual precipitation reaches 400–1200 mm a year, Mediterranean woodland and open parkland vegetation developed. Shrub land, steppic vegetation (Irano-Turanian), and desert plant associations (Saharo-Arabian) cover the areas where annual precipitation is less than 400 mm.

Presently, large annual rainfall fluctuations characterize the precipitation pattern in the Levant. The complicated climatic system (Wigley and Farmer 1982) makes it difficult to reconstruct the patterns of the past, but it is generally accepted that two annual patterns prevail. The first is controlled by the storm tracks which carry humidity from the Mediterranean Sea in a more arid southward direction; the second descends through Europe leaving most of the Southern Levant dry. Chemical studies of the Upper Pleistocene Lisan lake beds, which covered an area of about 2800 km^2 in the Jordan Valley, demonstrate that the geographic distribution of rainfall was similar to that of today (Begin et al. 1980). Fluctuations in decadal and centennial amounts of precipitation, rather than temperature changes, were responsible for expansion and contraction of vegetational belts as

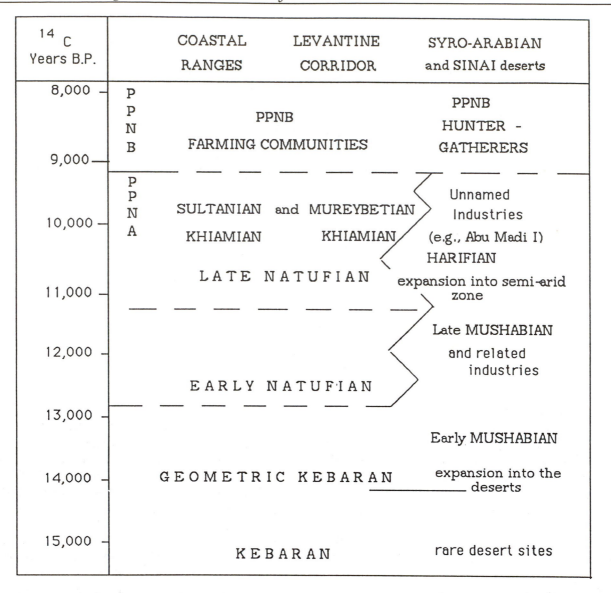

^{14}C Years B.P.	COASTAL RANGES	LEVANTINE CORRIDOR	SYRO-ARABIAN and SINAI deserts
	PPNB		PPNB
8,000 —		PPNB	
		FARMING COMMUNITIES	HUNTER - GATHERERS
9,000 —			
	PPNA		Unnamed Industries (e.g., Abu Madi I)
10,000 —	SULTANIAN and MUREYBETIAN KHIAMIAN KHIAMIAN		HARIFIAN expansion into semi-arid zone
11,000 —	LATE NATUFIAN		
12,000 —			Late MUSHABIAN and related industries
	EARLY NATUFIAN		
13,000 —			
14,000 —	GEOMETRIC KEBARAN		Early MUSHABIAN expansion into the deserts
15,000 —	KEBARAN		rare desert sites

Figure 1. A chronological chart indicating the names of the archaeological entities as used in the text and their distribution within the Levant.

reflected in the palynological sequences (Bottema and van Zeist 1981; van Zeist and Bottema 1982; Bottema 1987; Baruch and Bottema 1991).

The Paleo-Climatic Record

It is generally agreed that during the Last Glacial Maximum (LGM) the climate of the entire region was cold and dry but that the coastal hilly areas enjoyed winter precipitation and were covered by forests. This cold period was followed by a spell of wetter conditions 14,000–13,000 B.P., succeeded by an increase in arboreal pollen from about 13,500/13,000 B.P. which peaked around 11,500 B.P. (Bottema and van Zeist 1981; Baruch and Bottema 1991). The ensuing period,

beginning by 11,000 B.P., demonstrates a decrease in arboreal pollen (equivalent to the "Younger Dryas") and a return to pluvial conditions, although with a somewhat less marked peak around 10,000 B.P. A moister Early Holocene is therefore recorded both in the Hula basin as well as in the northern Levant.

Deep-sea cores in the Eastern Mediterranean (Nesteroff et al. 1983) clearly show that a short cold and dry spell separates the phase of Port Huron and the onset of the Bölling and Alleröd around 13,000/13,400 B.P. This sequence is supported by dated geomorphic events and paleosols (Goldberg 1986; Magaritz and Goodfriend 1987; Goodfriend 1991). Among the important changes was the main shrinkage of Lake Lisan from its maximum expansion at 180 m below sea level which occurred during the late Kebaran and before the Early Natufian (Bar-Yosef 1987b). The entire valley floor was exposed and the Jordan River, from its outlet in Lake Kineret, began incising a channel in the soft Lisan marls. The wetter and warmer period of the Early Natufian did not result in the full restoration of the lake. By the time of the PPNA, the site of Gesher, located in the Beith Shean (Beisan) Basin, is situated at 245 m below sea level. This evidence makes the suggested reconstruction of Lake Beisan (Koucky and Smith 1986) untenable.

Faunal spectra of Early Neolithic sites reflect the presence of a freshwater body in the area of the Lower Jordan Valley, indicating an increase in annual precipitation and especially the presence of copious springs (Tchernov in Noy et al. 1980; Leroi-Gourhan and Darmon 1987; Tchernov in Bar-Yosef et al. in press). Similar information about climatic amelioration around 10,000 B.P. was obtained at Mureybet (van Zeist 1988). Thus we may conclude that, at least during the terminal millennia of the Pleistocene and the early part of the Holocene, the entire Levant was subject to similar climatic fluctuations.

Finally, the sea rise since the Late Glacial Maximum was gradual and continued until the eighth millennium B.P. Over a period of about 7,000 radiocarbon years, the southern Levant with its flat, sandy coastal plain has lost a coastal stretch 10–15 km wide to the sea. In view of the paucity of aquatic resources in this most saline corner of the Mediterranean Sea, such a change affected only the collection of marine shells used for decoration.

The Seasonality of Vegetal and Animal Resources

Floral resources in the Levant are seasonal, with seeds most abundant in April-June and fruits in September-November (Schmida et al. 1986). Among the three vegetational zones, the Mediterranean zone is the richest in the number of edible fruits, seeds, leaves, tubers, etc. The fauna presents a similar picture, with biomass gradually declining away from the Mediterranean core area. Dense oak forests, where annual precipitation was above 800 mm, probably supported lower biomass than open parkland. Thus the mosaic associations of Mediterranean vegetation, between the isohyets of 600–300 mm bordering the Irano-Turanian shrub land, contain more floral and faunal species than other areas of the Levant (Shmida et al. 1986; Harrison 1968; Uerpmann 1981, 1987).

One of the main meat sources of Epi-Paleolithic and Early Neolithic communities was the gazelle (*Gazella gazella*). Recent studies of gazelle behavior (Baharav 1974, 1980, 1982, 1983; Simmons and Ilany 1975-1977)

indicate that this antelope is basically sedentary within a small home range. Females and fawns are more sedentary than the herds of bachelor males, which roam around territories controlled by dominant males. Thus the largest territories of the males in the most arid zone reach 25 km^2, decreasing considerably within the Mediterranean vegetation zone. Reproduction among gazelles depends on the daily accessibility of surface water. Under conditions that favor immediate availability of water, gazelles reproduce all year round. Drier conditions lead to seasonality in the timing of birth, generally March through July (Baharav 1983).

A somewhat similar pattern of behavior can be suggested for *Gazella subgutturosa*, the species commonly found in the Syro-Arabian desert (Harrison 1968; Meshel 1974 *contra* Legge and Rowley-Conwy 1987). Unfortunately, no behavioral studies of this species are available. Large herds of this gazelle subspecies were hunted with the aid of drives known in the Near East as "desert kites," beginning in the late ninth millennium B.P. (Meshel 1974; Bar-Yosef 1986; Legge and Rowley-Conwy 1987; Betts and Helms 1986).

Other common species, such as wild cattle (*Bos primigenius*), fallow deer (*Dama mesopotamica*), roe deer (*Capreolus capreolus*), wild boar (*Sus scrofa*), and ibex (*Capra ibex*) also have small home ranges like the gazelle. Thus, the reliability and accessibility of resources in the Mediterranean belt enable us to assume that the exploited vegetal and animal species were abundant and predictable. The absence of long-distance migrations of animals in the Levant did not require long-range residential or logistical moves. Resource exploitation along ascending or descending topographic cross-sections would have been the most fruitful.

Interpretations

The seasonality of vegetal resources would encourage movement between the lowlands and the highlands (which were always more forested in the Levant), while hunting, especially of gazelles, could be pursued within relatively small territories. Given the Levantine conditions, we have calculated the optimum exploitation territory for a band of hunter-gatherers within the Mediterranean vegetational belt to have been of the order of 300–500 km^2 (Bar-Yosef and Belfer-Cohen 1989).

Given the estimates related to the size of hunter-gatherer territories and the seasonal distribution of vegetal and faunal resources, we feel that the following predictions would explain the observed socioeconomic changes (Bar-Yosef and Belfer-Cohen

1989). If bands were confined to the specific Irano-Turanian and/or the desertic Saharo-Arabian phytogeographic zones, foraging territories would have been much larger than in the Mediterranean zone as a precaution against the hazards of annual fluctuations in resources. We estimate a territory of the order of 500–2000 km^2 for a single band. In a survey of the potential biomass of 500 km^2 in central Sinai, within the Saharo-Arabian belt (Perevolotsky and Baharav 1987), it was concluded that 10–20 people could survive solely by hunting; this figure would be higher if meat constituted only 10–20% of the diet, as is known from the ethnographic records.

Situations of food stress would be caused by dwindling resources as a result of decreasing annual precipitation and shifts in the distribution of rains during the winter, as well as by any increase of local population or by additional groups moving in. Climatic changes would affect largely the Irano-Turanian steppe and the Saharo-Arabian desert belts, while in the Mediterranean belt resources would have been more stable and fluctuations minimal (as explained above). A number of options would help to alleviate such stressful conditions. These are (a) population aggregation in core areas, which were least affected by the depletion of resources; (b) social and techno-economic reorganization within the same territories, in an effort to adapt to the new conditions; (c) immigration to adjacent territories, mainly northward (or westward for those who inhabited the Syro-Arabian desert), where conditions were still somewhat better due to their proximity to the Mediterranean zone; and (d) the combination of several operational strategies which under every circumstances require well-established social alliances.

Movement into the northern territories or westward along the Turkish coast would have been impractical because these areas were already occupied by local groups of hunter-gatherers (e.g., Albrecht 1988). The seasonal pattern of residential movements between winter base camps in the lowlands and summer camps in the highlands was most probably inherent within Mediterranean foragers' societies. However, it should be clear that this strategy of settlement does not necessarily mean that people did not use highlands in the winter and lowlands in the summer. For example, it was observed (Zohary 1969) that harvesting wild cereals would be profitable if it began in late May in the lowlands. By climbing the hilly slopes (such as between the Sea of Galilee and the town of Safad), gatherers would prolong the harvest period till the end of June or early July. Collection of wild barley, as known today from Ohalo II (Kislev, Nadel and Carmi in press) was already being practiced around 19,000 B.P. in this area. Regular patterns of band mobility would encourage the

building of stable installations, storage facilities, and solid foundations for dwellings, which could serve the same group during many successive seasons, enhancing the routes of anticipated moves.

More efficient exploitation would have been achieved by relocating the base camps on ecotones in order to intensively exploit a variety of environments. Large numbers of permanent storage facilities and increasing quantities of food refuse would attract rodents, birds, and scavengers. The continuous practice of a "broad-spectrum" exploitation strategy, which is essential for feeding a large and relatively stable group, would exert pressure on the game population, leading to depletion of certain species. The use of fire may have enhanced the growth of annuals and could even increase the annual yields of wild cereals and pulses (cf., Blumler and Byrne 1991). Thus the base camps situated in the ecotones became more sedentary settlements with increasing exploitation of wild legumes and cereals and the incentive to start some intentional cultivation.

The definition of sedentism is not a simple one (e.g., Rafferty 1985; Edwards, et al. 1988; Tangri and Wyncoll 1989; Tchernov 1991). There are numerous ethnographic examples from Southwest Asia of settlements with permanent houses and storage facilities from which most of the population moves out seasonally, often in summer time, either to fields in the lowlands or to the pastures in the highlands. When in need of particular objects, they return to the village and bring it back to the seasonal camp. The inhabitants consider themselves sedentary people and their settlements as the permanent dwelling of the group.

In order not to become entangled in a fruitless discussion, we adopt here the stance that the inference for sedentism versus mobility and short-term settlements should come from the biological evidence, accumulated from the excavated Epi-Paleolithic and Neolithic sites. We take into account the season during which the gazelle were hunted and the time of the year when seeds and fruits were collected. As the latter are often not preserved, we view the shift in the microvertebrate spectra and especially the dominance of house mice as an indication for a major change in mobility patterns and the establishment of sedentary communities (Tchernov 1984; Aufrey et al. 1988). Let us stress our viewpoint by stating that well-built structures and storage facilities alone do not indicate sedentism without supportive bio-archaeological evidence. Well-built structures were uncovered in Early Neolithic sites in southern Sinai, where the available information suggests seasonal residential moves similar to that of recent Bedouin (Bar-Yosef 1984; Edwards 1989). However, on the basis of the bio-archaeological contents of Natufian base camps, and plant remains as well as animal bones from Neolithic villages in the Jordan Valley as presented below, we infer that these were semi-sedentary or sedentary settlements.

Finally, the continuous exploitation of resources in a situation of population growth, when the retreat to a previous subsistence strategy is not feasible, would eventually lead to active human intervention to increase the yield of the exploited resources. This would be much easier to achieve with cereals and pulses because yields are expected within a few months. The shift from intensive collection to cultivation could not succeed under severe annual fluctuations of precipitation which characterize the Irano-Turanian steppic belt. The hazards of drought would not encourage the intentional sowing of about a quarter of the previous summer's yield. Therefore, cultivation would most likely have started in well-watered areas such as lake shores. The available paleoclimatic records from the Jordan Valley and the Damascus basin support this contention (e.g., van Zeist and Bakker-Heeres 1979, 1985).

The Geometric Kebaran and Mushabian Complexes

Geometric Kebaran—Observations

This entity exhibits techno-typological traits derived from the Levantine Kebaran (e.g., Bar-Yosef 1981b; Goring-Morris 1987; Cauvin 1981; Henry 1989), and is briefly described below. The age and duration of the Geometric Kebaran is based on a few stratified sites and numerous ^{14}C dates, mostly from the southern Levant, from about 14,500 B.P. to about 13,000/12,800 B.P. (Bar-Yosef and Vogel 1987).

Geometric Kebaran sites are dispersed in the Mediterranean belt as well as throughout the Negev, Sinai, and Syro-Jordanian deserts (Figure 2). Numerous locales in the arid zones are situated away from any perennial water sources known at present (Bar-Yosef and Belfer-Cohen 1989). Apparently, Geometric Kebaran hunter-gatherers took advantage of the climatic amelioration which occurred around 14,500-13,000 B.P. in the Levant and expanded into the semiarid zone.

Geometric Kebaran lithic assemblages are characterized by high frequencies of blades and bladelets, shaped primarily into microlithic trapeze-rectangles. It seems that narrow geometrics are a continuation of earlier Kebaran microlithic types, while the wider trapeze-rectangles represent a later stage when a proliferation in the production of blades took place. While the assemblages of the central Levant display a large variety of additional microlithic forms, the arid zone assemblages contain almost exclusively trapeze-

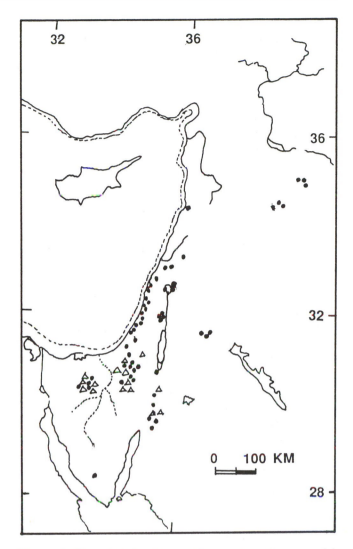

Figure 2. The distribution of Geometric Kebaran (•) and Mushabian sites (Δ) in the Levant (after Bar-Yosef and Belfer-Cohen 1989, Figure 3). Note that the paucity in the central and northern Levant reflects the state of research and not the real picture. The dashed lines mark the reconstructed shoreline of this period.

rectangles with a few exceptions (e.g., the "Nizzanan," Goring-Morris 1987, and see also Byrd and Garrard 1990). The final stage of the Geometric Kebaran in the semi-arid zone is marked by the introduction of backed lunates (Group IV of Henry 1989) and the appearance of a local facies in which the microlithic tool group is dominated by triangles as at Wadi Jilat 6 (Garrard et al. 1986; Garrard et al. 1987). Both industrial facies exhibit an intentional use of the micro-burin technique (Bar-Yosef 1975; Henry 1982; Goring-Morris 1987; Garrard et al. 1986; Byrd and Garrard 1990).

Pounding tools (pestles, bowls, and cup-holes) are usually found in Geometric Kebaran sites located within the Mediterranean belt (Bar-Yosef 1981b; Kaufman 1986).

The size of Geometric Kebaran sites is poorly known. Only a small number of sites has been excavated on a large scale. The size of the small sites is 15–25 m², others reach 100–150 m², and there are very few between 300 and 600 m² (Hours 1976; Kaufman 1986; Bar-Yosef 1975, Bar-Yosef 1981b). The small and medium-size sites in the Negev and Sinai may represent the remains of small, highly mobile bands, as indicated by the consistently similar lithic assemblages from sites widely separated in space.

Two fragmentary burials from the Geometric Kebaran were exposed from Neveh David, near the town of Haifa in the Mt. Carmel area (Kaufman 1986). Both skeletons are partially covered by stone bowls and mortars.

Marine shells such as *Dentalium* sp., *Columbella rustica*, and *Nassarius gibbosula* obtained from the Mediterranean sea shores were likely used for body and garment decorations (D.E. Bar-Yosef 1989). These shells have been found as far south as Wadi Feiran in southern Sinai (Bar-Yosef and Killbrew 1984).

Only scanty information about the economic activities of Geometric Kebaran hunter-gatherers is available. Animal bones are not preserved in the sandy areas of the Western Negev and northern Sinai. The main sources of meat in northern Israel were fallow deer, gazelle, and wild boar (Davis 1982; Bar-Yosef 1981a; Kaufman 1986), while in the Negev and Sinai we may assume that the prime game animals were gazelle, ibex, and hare.

The Mushabian Complex—Observations

The Mushabian complex is defined as a cultural entity on the basis of a set of specific techno-typological traits which characterize assemblages found in the northern Sinai and the Negev (Figure 2) as far north as the southern foothills of the Judean hills (Phillips and Minz 1977; Marks 1977; Goring-Morris 1987). It was suggested that the easternmost expansion is recorded in southern Transjordan (Henry 1983, 1989). The attribution of the Nahal Hadera sites, near Mt. Carmel, to this entity (Henry 1989) is untenable on the basis of techno-typological considerations.

The time range of the early phase of the Mushabian, based on stratigraphic evidence and radiocarbon dates, coincides with the Geometric Kebaran (14,500–12,800 B.P.) and in its late phase with the Early Natufian (12,800–11,000 B.P.) (Bar-Yosef and Belfer-Cohen 1989). The Mushabian lithic industry demonstrates affinities with the microlithic industries of Northeast Africa (Phillips 1973; Close 1978). A prominent characteristic of the reduction sequence is the intensive exploitation of the microburin technique, resulting in a high incidence of its products such as La Mouillah points. The combination of intensive use of microburin technique and the shapes of the microliths is clearly new to the Levant (Henry 1974). However, small scale use of this technique is already known from some Kebaran assemblages (e.g., Saxon et al. 1978) as well as from the Madamaghan (previously "Qalkhan" and "Late Hamran") assemblages (Henry 1983, 1989) dated to ca. 11,500 B.P. in the Azraq Basin (Garrard et al. 1986), and sites dominated by triangles near Ein Gev and in El-Kowm (Bar-Yosef 1985; Cauvin 1981).

The chronological position of the Late Mushabian (also known as "Ramonian," Goring-Morris 1987) is inferred mainly from its techno-typological properties and a few ^{14}C dates. The Late Mushabian contains Helwan lunates and is therefore considered to be contemporary with the Early Natufian (Marks and Simmons 1977; Bar-Yosef 1987a; Bar-Yosef and Belfer-Cohen 1989; Henry 1989). This chronological assignment implies that the Late Mushabian lasted until about 11,500/11,000 B.P. This conclusion is reinforced by the paucity of the Early Natufian sites in the Negev and their ephemeral nature as small flint surface scatters (Goring-Morris 1987).

Rare faunal remains from Mushabian sites and a few pounding tools can hardly reflect Mushabian economy, apart from the general assumption that they were hunter-gatherers exploiting the semi-arid steppic Irano-Turanian and the marginal areas of the Saharo-Arabian belts. It is worth noting that the extensive surveys in Saudi Arabia made it clear that these Epi-Paleolithic complexes are not present and should therefore be treated only as Levantine entities (Zarins 1990).

Additional archaeological assemblages, contemporary with the Mushabian, have been discovered and described in the Azraq Basin and in southern Jordan (Garrard et al. 1986; Muheisin 1985, 1988; Byrd and Garrard 1990).

Interpretations

Understanding the Geometric Kebaran and Mushabian is crucial, in our view, for those who would like to trace the prehistoric social groups that became Natufian around 12,800 B.P. The "impoverished" tool kits of the desert sites may indicate a specialized adaptation to specific semi-arid conditions which enforced a higher mobility compared to the potential for sedentism within the Mediterranean vegetational belt.

Identification of prehistoric social groups has been done on the basis of the typological variability within the Geometric Kebaran. In a recent essay, Henry (1989) interprets this variability as the result of the presence of four groups of assemblages:

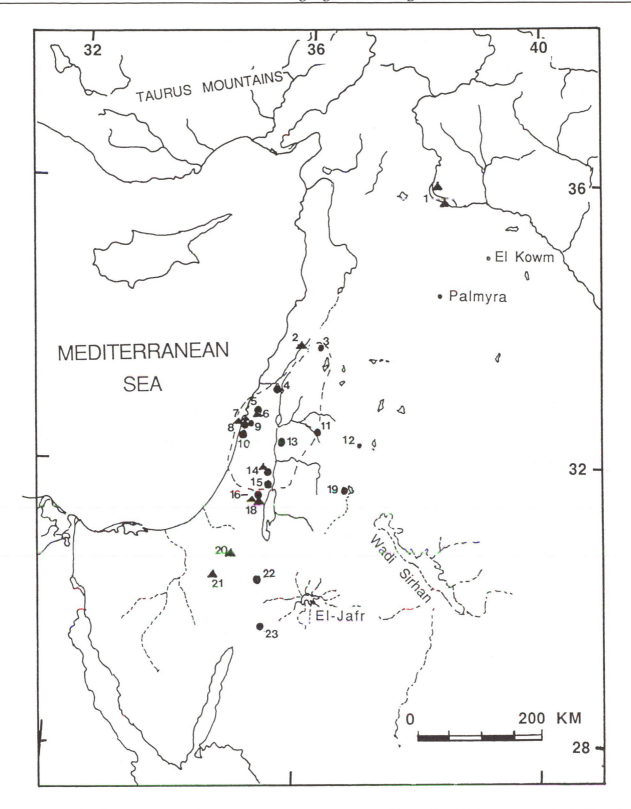

Figure 3. The distribution of most Natufian sites across the Levant. The circles (•) designate an Early Natufian occupation and the triangles (▲) designate the Late Natufian. (1) Mureybet and Abu Hureyra; (2) Saidié; (3) Yabrud III; (4) Ain Mallaha (Eynan); (5) Hayonim cave and terrace; (6) Hilazon; (7) Nahal Oren; (8) El-Wad cave and terrace; (9) Rakefet cave; (10) Kebra cave; (11) Taibé; (12) Khallat 'Azana; (13) Wadi Hammeh 27; (14) Fazael IV; (15) Salibiya I, XII, Gigal; (16-18) Erq el Ahmar, Umm Qala, Ain Sakhri, El-Khiam terrace; (19) Azraq 18; (20) Rosh Zin; (21) Rosh Horesha; (22) Beidha; (23) Wadi Judayid. The dashed line marks the approximate limits of the Natufian "homeland."

(a) Group I sites are confined to the Mediterranean core area and are characterized by the predominance of trapeze-rectangles (Henry 1989: 176). This definition is based on detailed inventories of five sites, three of which are located outside this vegetational belt. The sites attributed to this unit contain pounding and some grinding tools. Group I is considered to be contemporary with Group II.

(b) Group II sites (in the Negev, Sinai, southern Jordan and the Syrian desert) reflect an arid zone adaptation. Henry (1989) states that the sites were abandoned after 13,000 B.P., leaving the desert region to be occupied by Mushabian groups. Henry suggests that this group gave rise to Group IV.

(c) Group III sites, located in the coastal plain and Jordan Valley, are characterized by the proliferation of triangles (formerly called "Geometric Kebaran A2," Bar-Yosef 1975).

(d) Group IV demonstrates the greatest similarities to the Natufian in the lithics, expressed in the high frequencies of microburin technique and backed lunates (Henry 1989: 164). According to Henry the Natufian emerged from Group IV along the margins of the Mediterranean belt. This conclusion is based unfortunately on his lumping the entire array of Natufian assemblages into one entity and ignoring, for this purpose, temporal and spatial variability.

The reader should be reminded that the earliest Natufian assemblages, dated to 12,900–12,400 B.P. are in the Galilee and Mt. Carmel areas. These have none or low frequencies of by-products of the microburin technique and a dominance of Helwan lunates among the geometrics (e.g., Bar-Yosef and Valla 1979; Henry 1989:110).

The small size of most of the Geometric Kebaran sites is similar to that of Kebaran sites (Bar-Yosef and Belfer-Cohen 1989) and is interpreted as a reflection of small and mobile bands. The large site of Neveh David, at the foot of Mt. Carmel, may represent a size increase at favorable locations (Kaufman 1986).

The amazing techno-typological uniformity of lithic assemblages and site size in the arid region of the Negev and Sinai probably reflects a higher degree of mobility practiced by fewer bands than in the Mediterranean belt. The expansion of the Geometric Kebaran into the arid zone, while retaining the small size of their sites, may indicate population growth when compared to the preceding Kebaran, in response to greater abundance and/or predictability of food

resources in these areas. Under conditions of climatic improvement, when the Mediterranean vegetational belt would have expanded, we expect that mechanisms controlling the size of local populations would have been relaxed. Population growth, triggered by the addition of stable, reliable food resources, could lead to the exploitation of additional territory by the same group, or by the budding-off of new groups. Moreover, the continuous and widespread use of Mediterranean marine shells by the Geometric Kebarans, points to the preservation of social ties with their original homeland.

The opening of new territories for traditional exploitation strategies tempted contemporary bands of foragers, the Early Mushabians, originating in the Nile Valley, to expand into the Sinai (14,500–12,800 B.P.). The size of the Mushabian sites indicates small groups, similar to those of the Geometric Kebarans. Sporadic pounding tools may reflect the use of plant food but the poor preservation of bone and plant remains in the excavated sites prevents a better understanding of their foraging strategies. If the North African origin is accepted, then they were simply expanding the territories into a region which resembled their ancestral habitat. The Geometric Kebarans, on the other hand, adapted themselves to a somewhat different biotope by moving from the Mediterranean to the Irano-Turanian and Saharo-Arabian belts. Perhaps this contrast between the two populations explains their different rates of success.

The radiometric contemporaneity of the Geometric Kebaran and the Early Mushabian complex led Henry (1989:146) to suggest that the expanding Geometric Kebarans (Group II) were influenced by the Mushabians, the occupants of the arid belt, and that there was a uni-directional cultural diffusion when the Geometric Kebaran (Group IV) borrowed tool forms and techniques from the Mushabian. The major socio-economic changes took place as early as the thirteenth millennium B.P. The Geometric Kebarans disappeared from the arid south and the Natufian cultural complex was established. There are no Geometric Kebaran sites contemporary with Late Mushabian sites, which are contemporary with the Early Natufian. The small ephemeral Mushabian sites in the lowlands and a few larger ones in the Negev highlands reflect the mobile hunting and gathering way of life of desert groups that survived while the Natufian settlement pattern prevailed in the more lush zone.

Natufian Complex

Observations

Sites classified as Natufian by various archaeologists are found throughout the Levant, from the middle Euphrates to the Negev highlands and along the Jordanian plateau (Perrot 1968; Braidwood 1975; Bar-Yosef 1983; Henry 1985, 1989; Bar-Yosef and Belfer-Cohen 1989; Belfer-Cohen 1989, Belfer-Cohen et al. 1991). While this geographically broad definition may be correct, the Natufian "homeland" seems to have been in the central Levant. Radiometric dates indicate that the chronological boundaries of the Natufian are ca. 12,800–10,500 B.P. (Valla 1987; Bar-Yosef and Belfer-Cohen 1989). However, a subdivision into Early and Late Natufian is of crucial importance. Most of the simplified descriptions of the Natufian tend to lump both phases together (e.g., McCorriston and Hole 1991). We have tried to demonstrate in previous papers (Bar-Yosef and Belfer-Cohen 1989; Belfer-Cohen 1989, 1991) that there are major differences between the two phases that were likely the result of significant social changes. This issue will be discussed further below.

We begin by presenting the characteristics of the Natufian as a socio-economic unit, which within its original "homeland" in the central Levant (Figure 3), represents what is currently interpreted as sedentary hunter-gatherers and therefore different from their predecessors.

Natufian sites fall into three size categories: small (15-100 m^2), medium (400-500 m^2) and large (more than 1000m^2) (e.g., Bar-Yosef 1983; Byrd 1989; Bar-Yosef and Belfer-Cohen 1989). Only small Early Natufian sites were documented in the Irano-Turanian zone while larger sites in this belt are solely of Late Natufian age (e.g., Betts 1982; Henry 1982, 1983; Moore 1985; Goring-Morris 1987).

Cave occupations are an interesting aspect of the Natufian settlement pattern rarely encountered during Kebaran and Geometric Kebaran times. The Natufians reoccupied most of the caves and rock shelters that had been inhabited for short time spans during the Upper Paleolithic and were later abandoned. The caves were generally dry and could be used for human occupation and storage (Bar-Yosef and Martin 1979). The well-preserved dwelling structures at Mallaha and Wadi Hammeh 27 provide an idea about the building techniques and forms of Early Natufian structures including some storage facilities (Perrot 1966; Valla 1981, 1984; Edwards et al. 1988; Bar-Yosef and Goren 1973; Belfer-Cohen 1988b). The structures are rounded and their diameter vary from two to nine meters (Figures 4-5). Hearths were uncovered nearly in every building. Evidence for the use of plaster and several post-holes uncovered in a large house at Mallaha indicates the use of wood for roofing (Valla 1988). Fragmentary stone walls were documented at El-Wad Terrace and Nahal Oren. It should be stressed that most of the known well-built dwellings are dated to

the Early and Middle Natufian by Valla's definitions (Valla 1987).

The density of lithics per cubic meter in Natufian sites and the exploitation of different types of flint and limestone indicate intensive occupations. Natufian core reduction techniques resulted in high frequencies of flakes and broad, short bladelets. Many of the latter were shaped into microliths, dominated by lunates. The presence/absence of microburin technique, as well as variability among the lunates can be used as criteria for subdividing the Natufian into phases and regional groups (Bar-Yosef and Valla 1979). The average length of lunates has also been used as a chronological marker (Valla 1984) and was recently refined to include the regional-ecological location of the sites (Olszewski 1986). On the whole, typological variability within the microlithic tool group is greater in Early Natufian sites than in Late Natufian ones.

The frequencies of various tool groups in Natufian assemblages vary according to the location of the site (Bar-Yosef and Belfer-Cohen 1989; Byrd 1989). High frequencies of sickle blades characterize Natufian sites situated in the "homeland" area, while low frequencies are noted at Mureybet and Abu Hureyra (Olszewski 1989). Both the sickle blades and the elongated picks appear for the first time in the Natufian and are the forerunners of succeeding Neolithic tool kits (Perrot 1966; Bar-Yosef 1983; Valla 1984). Other tool types such as retouched notches and denticulates, as well as borers and awls, are found in nearly every Natufian assemblage.

The Natufian bone industry is unique in its richness, variability, and decoration. Mundane, domestic bone tools designed for hunting and fishing, hide-working, and basketry are commonly found (Bar-Yosef and Tchernov 1967; Stordeur 1981, 1988; Belfer-Cohen 1991). The spatial distribution of these artifacts also suggests geographical subdivisions within the Natufian. Sites in the Mt. Carmel-Galilee area (and probably northern Jordan) have the richest, most varied collections (Stordeur 1991).

Groundstone tools are made of limestone, basalt, or sandstone. The primary types are pounding tools, but there are also mullers, whetstones, heavy duty scrapers, shaft straighteners, hammerstones, etc. A unique type are the large limestone "stone pipes" which may have been deep mortars; these artifacts were usually breached through and in several instances been placed vertically in graves (Nahal Oren), around a platform (Jericho), or near dwellings (Hayonim Terrace) (Stekelis and Yizraeli 1963; Kenyon 1981; Valla et al. 1989). The specific origin of the basalt and sandstone is yet not properly investigated. In most cases the Natufian base camps are at least 30 km away from the nearest basalt sources. It seems that the arti-

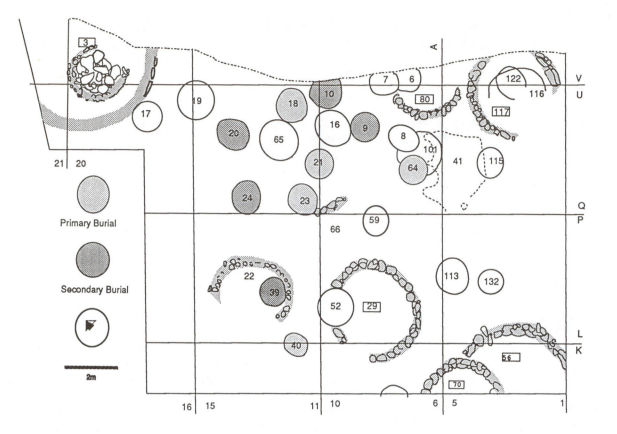

Figure 4. Schematic plan of layer Ic at Ain Mallaha (after Perrot and Ladiray 1988). (1) dwellings ("pithouses"); (2) pits (storage?); (3) graves (both primary and secondary burials).

facts were usually brought to the sites as finished products.

Natufian jewelery and decorative elements are varied and numerous. Beads and pendants were made of limestone, basalt, greenstone, malachite, bone, tooth, and a great variety of marine mollusks, especially *Dentalium* shells. Exotic materials testify to connections with neighboring regions. At Mallaha, obsidian from Anatolia and the freshwater shell *Aspathria* from the Nile are unique finds (Valla 1987). Greenstone beads were found at many sites and are thought to originate in Syria, Jordan, or the Sinai. Marine shells are commonly from the Mediterranean but a few were brought from the Red Sea (Mienis 1987; Bar-Yosef 1989).

Art objects are an additional archaeological characteristic which emphasizes the uniqueness of the Natufian among the Epi-Paleolithic cultures (Belfer-Cohen 1988a). Animal figurines, often interpreted as young gazelles (e.g., Cauvin 1972) or young ungulates, were carved from stone or bone. A few human representations herald what will become a major subject in the Neolithic (Cauvin 1972; Bar-Yosef 1983). Several limestone slabs with incised geometric forms and one large fish (?) were found in Hayonim Cave (Belfer-Cohen 1988b, Belfer-Cohen 1991). Larger carved limestone slabs, exhibiting a meander pattern, were uncovered at Wadi Hammeh 27 (Edwards et al. 1988).

Natufian burials are diversified in position (flexed, semiflexed, and extended), number of individuals per grave (ranging from one to five or more), grave structure, and decorations (found only in Early Natufian graves). The first evidence for selective skull removal was observed among Late Natufian burials at Hayonim cave (Belfer-Cohen 1989). The studied skeletons belong to the Proto-Mediterranean stock (Arensburg and Rak 1979). Burials of children comprise about one-third of the dead, with a relatively high mortality among those aged 5–7 years (Belfer-Cohen et al. 1991).

Late Natufian Desertic Adaptations

The Late Natufian, as opposed to the early stage, extended over a larger geographic region, from the middle Euphrates Valley (Figure 4), where it was discovered at Mureybet and Abu Hureyra (Cauvin 1977; Olszewski 1986) to the Negev. Two major sites were excavated in the Negev highlands: Rosh Zin and Rosh Horesha (Henry 1976; Marks and Larson 1977; Goring-Morris 1987). In the lowlands of the western Negev and northern Sinai small sites were found containing a typologically limited tool kit, including lunates, endscrapers, retouched blades, etc., with intensive use of the microburin technique.

A seasonal settlement pattern can be reconstructed on the basis of indirect evidence including the differences between summer and winter temperatures, location above sea level, and availability of water. Thus, small winter camp sites are well dispersed in the sandy lowlands, stretching from the seashore to about 60 km inland. These sandy areas have a much higher carrying capacity than the loessic plains. The summer or early spring marked the movement into the highlands. Local food resources were exploited, including hunting of the gazelle, ibex, and rabbits. Intensive collecting of vegetal food stuffs is indirectly evidenced by the presence of numerous pounding tools including bedrock mortars and some grinding stones. Compared to Mediterranean sites, base camps of the Late Natufian in the Negev area lack burials, building activities, art objects, and a rich bone industry. Flimsy structures were uncovered at Rosh Zin and Rosh Horesha (Marks and Larson 1977; for a different view of the latter site see Goring-Morris 1987). A great variety of sea shells, especially of Mediterranean origin as well as those from the Red Sea, point to the development of an exchange network among desertic groups or to the aggregation of groups which, during the winter, descended into the Arava Valley and the Gulf of Eilat.

During the eleventh millennium B.P. (ca. 10,800–10,300/200 B.P.) a phase of increasing aridity is recorded, correlated with the "Younger Dryas" interval (Magaritz and Goodfriend 1987). The archaeological unit which occupies this timespan in the Negev and Northern Sinai is known as the Harifian and is considered a cultural continuation from the Late Natufian.

Harifian settlement pattern reflects an adaptation to the local conditions of increasing aridity and the timing of harvests for wild barley and collecting pistachio nuts. Summer aggregation base camps are located on the Har Harif plateau and its surroundings

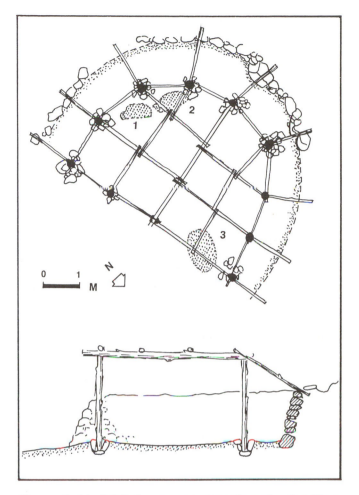

Figure 5. Ain Mallaha, a reconstruction of a dwelling structure from layer (after Valla 1988). 1, 2, 3, are hearths.

(900–1000 m above sea level) and transitory, usually small, winter camp sites are located in the sandy lowlands of the western Negev and northern Sinai. The known size of the Harifian territory is about 8,000 km², based on surveyed area, but it may have been 30,000 km² or more including most of the Negev and northern Sinai (Bar-Yosef and Belfer-Cohen 1989).

Finds from the small sites are mainly lithics, including lunates, endscrapers, and Harif points shaped by the microburin technique. Differences in the frequencies of tools among the various assemblages may indicate different discard patterns characterizing task specific sites.

The larger aggregation sites on Har Harif display a variety of remains (Goring-Morris 1987, 1991). Dwelling and storage structures have been uncovered, usually dug into the loess. Dwelling structures are often three meters in diameter. Inside and nearby are numerous limestone slabs with cup-marks, as well as a few mortars, pestles, and grinding stones. On the rocky slopes near the site of Rosh Horesha, several dozen cup-holes were found. Similar concentrations are known from the Judean desert and the site of Hatoula on the western edge of the Judean hills (LeChevallier and Ronen 1985).

The lithic industry of the Harifian resembles the Late Natufian in the dominance of microliths and especially backed lunates, as well as the intensive use of microburin technique. A special projectile point, mentioned above, is the Harif point, which occurs in low frequencies in the sites on Har Harif and higher frequencies in the lowland occupations (Goring-Morris 1987).

Vegetal material was not preserved but on the basis of the present flora, it has been suggested that the food resources included wild barley, legumes, and nuts. Meat was obtained through hunting of ibex, gazelle, rabbits, and perhaps also wild sheep (Davis et al. 1981).

Marine shell assemblages are somewhat different from those of the Late Natufian. Two-thirds of the species originated in the Red Sea and only one-third came from the Mediterranean (Mienis 1977).

Despite intensive surveys, no archaeological continuity for the Harifian was found (Goring-Morris 1987). Sites dated to the following tenth millennium B.P. are very rare. It seems that the efforts of the Harifians, who used improved traditional methods of hunting (Harif projectile points) and gathering to survive in territories adjacent to sedentary food-producing communities, have failed.

Interpretations

The emergence of the Natufian is a crucial subject because the Natufian is seen as a threshold for the establishment of farming communities. The conflicting views of gradual process versus sudden change are indirectly reflected in the literature. Henry (1989: 180) views the Natufian as a sudden socioeconomic shift but does not offer an explanation other than the availability of cereals as a prerequisite. Hole (1984), Henry (1989), and McCorriston and Hole (1991) base their interpretations on a previous suggestion by Wright (1977) who mistakenly concluded that wild cereals were generally not available in the Near East during the closing millennia of the Pleistocene. The recent discovery of numerous carbonized plant remains in Ohalo II, a 19,000 B.P. water-logged site in the Sea of Galilee (Kislev et al. in press; Nadel and Hershkovitz in press), reinforces the interpretation (based on scanty remains, such as the 30,000 year old grain in Nahal Oren, a few pieces of wild barley in the Natufian at Hayonim Cave, etc.) that cereals were indeed present in the Levant. Statements based on the paucity or lack of plant remains due to problems of recovery and/or preservation are definitely misleading. The contention

that pounding tools such as mortars and pestles may have been used for processing acorns, nuts, cereals, dry legumes, red ochre, and burnt limestone can be maintained in view of the available plant remains from Ohalo II. Furthermore, a study of past environments in the Jordan Valley (Nadel and Hershkovitz in press) demonstrates that similar plant communities to those of today were present in this region at the end of the Pleistocene and beginning of the Early Holocene.

It is not known whether the various aspects of the Natufian socioeconomic system (such as sedentism, cereal exploitation, etc.) are causally interrelated or not. For example, McCorriston and Hole (1991) suggest that sedentism enhanced the propagation of annuals such as cereals, while Henry (1989) advocates the reverse, namely that intensification of cereal exploitation brought about the Natufian sedentism.

Whether storage was practiced is not well known, but it is conceivable that the use of baskets (as evidenced by use wear on bone tools, see Campana 1989) began in the Natufian, if not earlier, and allowed the transport and storing of surplus.

The use of sickles seems to be of crucial importance. It is obvious from ethnographic evidence as well as experimental studies (Hillman and Davies 1990a) that sickles are more awkward to use than baskets and beaters. However, harvesting with sickles is more efficient in terms of energy expenditure and size of plots. It is therefore not surprising that the "homeland" Natufian assemblages contain many more sickle blades than those of sites like Mureybet or Abu Hureyra I where territories were larger and demographic pressure was lower. Thus we see the use of sickles as another indicator of increased territorialism during the Natufian.

The size and contents of Natufian sites reflect somewhat larger bands than those of the preceding periods. One of the largest Natufian sites, Mallaha (Eynan) in the Hula Valley, has been classified as a village (Perrot 1966; Cauvin 1978; Valla 1981). The investment in leveling slopes in order to build houses on terraces, the production of plaster, the transportation of heavy undressed stones into open-air and cave sites (e.g., Hayonim Cave), and the digging of underground storage pits (Mallaha) indicate energy expenditure expected at base camps but not anticipated for ephemeral, short-season occupations. However, it is the faunal and not the architectural evidence which, in our view, documents continuous seasonal or permanent residence in these larger sites. For the first time, human commensals (house mouse, rat, and house sparrow, which are self-domesticating species) are found and in large numbers (Aufrey et al. 1988; Tchernov 1991). This increase in the length of habitation is supported by the frequencies of immature gazelle bones as well as the study of gazelle tooth increments (Lieberman et al. 1990) which suggest year-round hunting (Davis 1983).

While the first threshold in the process that led to the emergence of agriculture was the establishment of the Natufian culture in its homeland in the central Levant, the second crisis came with the environmental deterioration imposed by the "Younger Dryas." This short climatic phase, which probably lasted several centuries, had a different impact on the Late Natufian population in Negev/northern Sinai region when compared to the impact on mobile hunter-gatherers who occupied the northern latitudes in Eurasia. This environmental change, while the Natufian "homeland" was inhabited by sedentary communities, forced the Late Natufians in the Negev to survive mainly on the exploitation of their local food resources. Efforts to improve their hunting methods brought about the invention of a special arrowhead named the "Harif point," which may also indicate an improvement in the bow (Bar-Yosef 1987a). However, in spite of these efforts the Harifian entity disappeared. It is unknown whether these people died out or ultimately joined their Late Natufian relatives in other areas in newly established Early Neolithic communities. Their disappearance left the Negev literally unoccupied for several hundred years.

The Early Neolithic

Terminology

The term "Neolithic" was originally used in European prehistory to designate assemblages with pottery, polished axes, domesticated cereals, and domesticated animals. Sites were often settled villages which differed considerably from the preceding Mesolithic ones. The Neolithic as a concept underwent considerable change in the Near East. Braidwood and his associates introduced a new descriptive nomenclature that was based on the socioeconomic interpretation of the available sites (Braidwood and Braidwood 1953). Kenyon, while digging in Jericho, suggested the term "Pre Pottery Neolithic" (subdivided into A and B), for the pre-ceramic levels in Jericho (Kenyon 1957). The state of knowledge concerning plant and animal domestication at that point was rather poor. A similar classificatory system, based on subdivision of time spans, such as period 1, 2, 3... was suggested by the Lyon school (Aurenche et al. 1981). Kenyon's terms are still commonly used in the Levant in their abbreviated form "PPNA" and "PPNB" (Figure 1).

The Khiamian—Observations

We maintain the approach that distinct archaeological entities can be identified through the techno-typological study of their lithic assemblages and we have followed Crowfoot-Payne (Crowfoot-Payne 1976, 1983) by distinguishing the "Khiamian" as a transitional entity seemingly contemporary with Mureybet phase IB and II (Cauvin 1977, 1978, 1990) and earlier or partially contemporary with the "Sultanian", which is a full-blown early Neolithic farming society (Figure 6).

The Khiamian is still an ill-defined unit, in our view, for two reasons: (a) the short time span of its existence (probably 200 or 300 radiocarbon years, ca. 10,500–10,300/10,100 B.P.) which causes overlapping radiocarbon readings from Late Natufian or Early Neolithic sites to confuse archaeologists who are looking for a clear-cut picture based on ^{14}C dates; and (b) the available information is derived from very limited soundings or from sites where admixture with earlier layers is likely (Echegaray 1966; Bar-Yosef 1980, 1989; LeChevallier and Ronen 1985; Cauvin 1978, 1989). This unit encompasses the stage Mureybet I B, previously called "Epi-Natufian" (Cauvin 1977) and Mureybet II (Cauvin 1990). The stratigraphic sequence at Mureybet supports the identification of the Khiamian as a separate entity from the Mureybetian (in the Euphrates River Valley) and the Sultanian in the southern Levant. However, consistent with our typological approach we do not include sites like Abu Madi I in this entity although it dates more or less to the same time span (Bar-Yosef 1991). Finally, lumping the Khiamian and the Sultanian as one cultural entity, as recently suggested (Garfinkel and Nadel 1989) is not advisable. By keeping each as a separate entity we may be able in the near future to better define the socioeconomic changes which occurred around 10,300–10,000 B.P. It is sufficient for classificatory purposes that both entities, and the desertic unit represent by Abu Madi I are all incorporated into the PPNA period.

The main characteristics of the Khiamian industry are the Khiam arrowhead, the asphalt-hafted—often not retouched—sickle blade, low frequencies of microliths (often including lunates), perforators, and the lack of bifacial or polished celts. Common core reduction strategies were aimed at producing blades and flakes, with a pronounced decrease in the frequency of bladelets. The small sounding in Salibiya IX produced faunal remains similar to those from Netiv Hagdud (a nearby Sultanian site) and a few carbonized seeds of pulses and some wild barley (Kislev, personal communication).

Another site dated to about the same period (10,100–9,800 B.P.) is Abu Madi I which lies at an elevation of 1600 m in Southern Sinai. It is considered as a summer occupation given the known harsh climatic conditions of the winter in this region (Bar-Yosef 1985). It contains a well-built, subterranean house, 4 m in diameter; a bell-shaped storage pit has been uncovered outside the house. The hunted animals were mostly ibex, with some gazelle and hare. Dried fish may have been brought from the Red Sea as well as a large collection of sea shells.

The dominant arrowheads were Khiam points which characterize assemblages dated to the end of the eleventh and most of the tenth millennium B.P from northern Iraq (Watkins et al. 1989) through southern Sinai. A local projectile point, with a foliate shape and sometimes a tang, named the Abu Madi type, was found in similar frequencies (Bar-Yosef 1985). Among the microliths, both backed and Helwan lunates were present. The latter are considered an anachronism compared to the Mediterranean Levant, where they disappeared at least 700 radiocarbon years earlier. Grinding and pounding tools at this site include handstones, flat grinding slabs, and a few pestles.

The Sultanian and the Mureybetian– Observations

The Sultanian, dated to ca. 10,300/100–9,300/200 B.P. by numerous radiocarbon dates (Bar-Yosef 1991) is known from the excavations of Jericho (Kenyon 1981; Kenyon and Holland 1983), Gilgal (Noy 1989), Netiv Hagdud (Bar-Yosef et al. in press), Gesher (Garfinkel and Nadel 1989), Hatoula (Lechevallier and Ronen 1985), Nahal Oren (Stekelis and Yizraeli 1963) and Iraq Ed-Dubb, a cave site on the western flanks of the Transjordanian Plateau (Kuijt et al. 1991) (Figure 6).

Jericho, Gilgal, and Netiv Hagdud are mounds which contain the remains of rounded and oval houses (Figure 7). The houses are semi-subterranean, with stone-lined foundations and superstructures built of unbaked loaf-shaped or plano-convex mudbricks. The use of mudbricks along with considerable amounts of organic substances caused rapid accumulation in these Neolithic mounds and low artifact frequencies per excavated volume, compared to Natufian sites.

Communal activities are evident in the walls and the tower of Jericho, interpreted by Kenyon as part of a defensive system (Kenyon 1957). An alternative interpretation (Bar-Yosef 1986) suggests that the walls, built up in various thicknesses, were erected as a protection for the settlement against mud flows and flash floods. The tower, built inside the walls, could accommodate on its top a small mud-brick shrine. Evidence for unequivocal public ritual, however, is still missing, probably due to the small area excavated in these sites.

Hearths were found inside the houses and in open spaces. One type of hearth at Netiv Hagdud was an oval basin-like floor of cobbles. Cooking involved heated stones which created a proliferation of "firecracked rocks" again unmatched in the Natufian.

The lithic technology shows a slight preference for blades, but often the existing cores do not match the available products. The variety of flint types indicates raw material procurement from diverse sources and possibly from greater distances than previously. There is also evidence for lithic heat treatment. The lithic industry is characterized by Khiam arrowheads, perforators (on blades and flakes), various types of sickle blades including a type with bifacial edge retouch (the Beit Ta'amir knife), axes with an edge formed by a transverse blow ("Tahunian" celts), burins, and a few scrapers. Polished celts made from limestone and basalt occur as well. Pounding tools include slabs with cup-marks (or cup-holes), numerous pestles, shallow grinding bowls, and hand rubbers (*manos*).

The assemblages of Mureybet III and Tel Aswad I A (Cauvin 1974; Cauvin and Stordeur 1978; Cauvin 1989) exhibit regional differences in certain lithic types including a *herminette* (an adze made on a thick flake), perforators, relatively high frequencies of scrapers and burins, decreasing frequencies of Khiam points and increasing frequencies of Helwan points (tanged with bi-lateral notches). It seems that both forms of projectile points are common denominators across the Levant and that the latter appears first in the north and later in the south (Gopher 1989).

Long-distance exchange is evidenced by central Anatolian obsidian which occurs at Jericho, in small quantities in Netiv Hagdud, and as a few pieces in Nahal Oren and Hatoula; none was found in Gilgal or Gesher. Marine shells were brought from the Mediterranean coast and less often from the Red Sea. There is a clear shift in the type of shells selected for exchange. *Glycymeris* and cowries became important but *Dentalium* shells (where excavated deposits have been sieved) were still common as in Natufian sites (D.E. Bar-Yosef 1989).

Site size presents a clear hierarchy from large to small: Mureybet, Tel Aswad, and Jericho (5.0–2.5 ha); Netiv Hagdud and Tel Aswad (1.5–1.0 ha); Gilgal (1.0–0.5 ha); and Nahal Oren and Hatula (0.5–0.2 ha). Preliminary observations of intrasite variability are based on the excavations at Nahal Oren and Netiv Hagdud. At Nahal Oren, the houses are almost all of the same size and clustered together resembling the Natufian settlement at Mallaha, while at Netiv Hagdud houses are clearly of different sizes and have larger open spaces between them. However, much more published field evidence is required before any conclusions concerning the size and wealth of the various households can be reached.

Burials and art objects shed a little light on beliefs. Most adult burials in Jericho and Netiv Hagdud are single, lacking grave goods and often without skulls. The crania were removed one or two years after death,

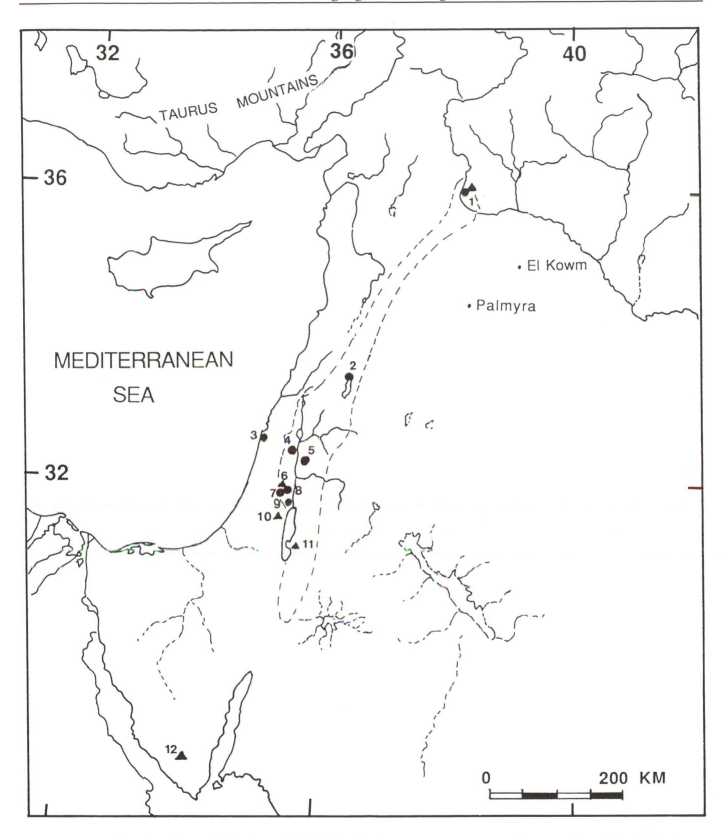

Figure 6. The distribution of selected Early Neolithic and contemporary sites, mentioned in the text. (1) Mureybet; (2) Tel Aswad; (3) Nahal Oren; (4) Gesher; (5) Iraq ed Dubb cave; (6) Salibiya IX; (7) Netiv Hagdud; (8) Gigal; (9) Jericho; (10) El Khiam terrace; (11) Dra'; (12) Abu Madi I.

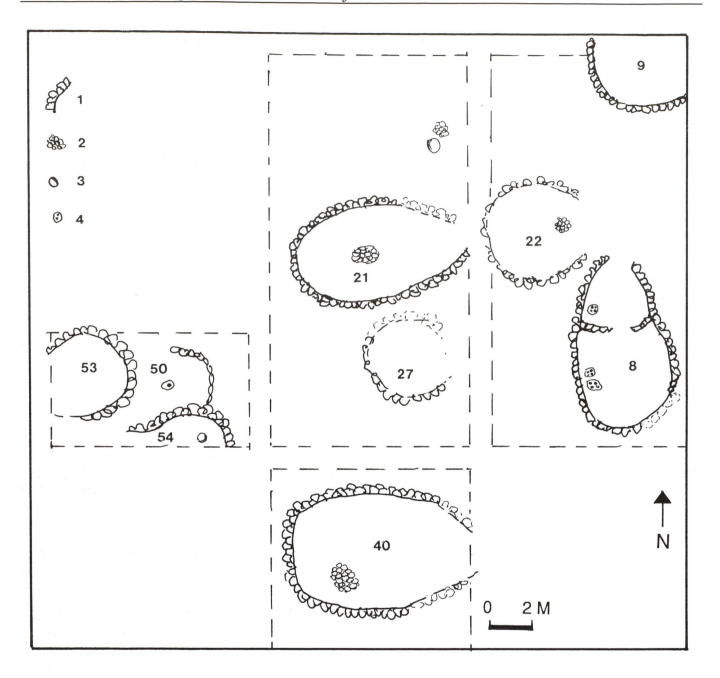

Figure 7. Schematic plan of Netiv Hagdud (after Bar-Yosef et al. in press). The upper right rectangle and the lower square represent a layer later than the remains in the other rectangles: (1) walls of dwellings 'pit houses'); (2) fireplaces; (3) grinding bowls; (4) slab with cup-holes.

and the lower jaws were left in place. No plastered or decorated skulls have been found as yet, although the technology for making lime plaster (and gypsum plaster in the northern Levant) was known. At Netiv Hagdud a cluster of a few crushed skulls was uncovered on the floor of a house which contained three slabs with cup-holes and numerous grinding and pounding tools.

Art objects are few (Bar-Yosef 1980). Those carved of stone from Salibiya IX, Gilgal (surface), and Nahal Oren are interpreted as depicting a schematic kneeling female figure. At Netiv Hagdud, three baked clay figurines of the "seated woman" were found. The Sultanian figurines, despite their rarity, indicate a shift from Natufian representations in which animals and indistinct human figures are more prominent. There is the appearance of female figurines, which marks, according to Cauvin, the cultural expression for the dichotomy between males and females and which plays a more important role in succeeding millennia (Cauvin 1985).

Most scholars agree that the economy of these Early Neolithic communities was based on the consumption of cereals and legumes, wild seeds and fruits, hunted mammals, and gathered reptiles, birds and fish (e.g., Clutton-Brock 1979; Tchernov in Noy et al. 1980; Davis 1983; Helmer 1989). Acquisition techniques varied. Identifying the ways in which early Neolithic communities, large and small, secured their food is therefore not only of great interest but also an issue of great debate.

The hunted game species, as a rule, are reflected in the bone counts although there are various ways for calculating their frequencies. Most game within the Mediterranean Levant were the same species. Gazelle is the dominant species at Mureybet, with wild ass (*Equus asinus*), rare cattle, caprovines, fallow deer, and wild boar also recovered. Jericho, on the other hand, is represented by a dominance of gazelle and a fair amount of fox with rare wild boar, deer, and cattle. Although sample size is smaller, Netiv Hagdud and Gilgal both resemble Jericho. However, these two sites where systematic sieving was practiced provided rich assemblages of bird bones, mostly of various species of ducks (Tchernov in Noy et al. 1980; Bar-Yosef, et al. in press).

At the site of Nahal Oren in Mt. Carmel, gazelle is dominant, along with a few forest animals such as the fallow and roe deer (Legge in Noy et al. 1973). At Hatoula, situated on the western slopes of the Judean foothills and at the same latitude as Jericho, gazelle, hare and fox are dominant with a few bones of aurochs, wild boar, equid, hedgehog, small carnivores, and marine fish (Davis 1983).

The overall picture that these remains provide is that hunting and trapping were the main techniques employed by the PPNA inhabitants of these sites. Whether the meat of certain animals was obtained by exchange with the neighboring hunter-gatherers is unknown. In view of the information from the following PPNB period (Bar-Yosef and Belfer-Cohen 1988), this is a viable option. The shift to the dominance of caprovines in the Levantine corridor, which we accept as the result of incipient domestication, occurred only during the mid-PPNB times (Bar-Yosef 1981a; Davis 1983; Helmer 1989; Byrd, this volume).

The second subsistence strategy, and probably the most important in terms of caloric intake, was the acquisition of vegetal food-stuffs such as a large variety of seeds and fruits, tubers and leaves (that are poorly represented in the archaeological record), and

legumes and cereals (van Zeist and Bakker-Heeres 1985, 1986; Hopf 1983; Kislev 1989, in press; Kislev et al. in press; Zohary and Hopf 1988; Hillman and Davies 1990a; Hillman et al. 1989).

One of the major questions centers on the shift from systematic gathering to cultivation. The best recorded pre-Neolithic floral assemblages have come from the Epi-Paleolithic site of Abu Hureyra on the middle Euphrates River (Hillman et al. 1989). Hillman and his associates identified 150 species of locally available modern seed plants in this area from May through July and from August to November. The gathering seasons do not necessarily reflect the time of consumption and thus the authors have argued that the site was occupied during most of the year. Importantly, there is no positive evidence to demonstrate that the wild einkorn wheat or wild rye were cultivated. SEM studies of the grains have not disclosed histological characteristics indicative of domestication. Furthermore, field observation in the steppe region in the wet winter of 1983 indicated that the distribution of "weedy" species was not limited only to cultivated fields (Hillman et al. 1989).

Unfortunately Abu Hureyra was not occupied during most of the PPNA period. Mureybet III (van Zeist and Bakker-Herres 1986) provided about 60 species of wild seeds and fruits, and a large number of two-seeded wild einkorn wheat grains, which were also found in earlier phases but became numerous in phase III. Size classes of lentil seeds indicate that they were also collected in the wild.

There is still considerable disagreement as to *when* the transition to planned cultivation occurred. In a recent survey of the available evidence for domestication, Hillman and Davies accept identifications based of grain morphology, size, and rare well preserved rachis fragments, as demonstrating the presence of domesticated wheat and barley in Tel Aswad, Jericho, Gilgal, and Netiv Hagdud (Hillman and Davies 1990b). Conversely, Kislev (1989, in press) observed that among threshed ears of wild barley 10% of the internodes still retain a fragment of the upper internode attached to the articulation scar. In some cases several internodes remain intact. Thus, the lower part of the ear of wild barley exhibits attributes that are considered at present as characteristic of domesticated forms. Kislev's conclusions are that most of the barley uncovered in Gilgal and Netiv Hagdud do not necessarily indicate intentional cultivation.

Zohary disagrees and in an experiment conducted in Jalés (Ardeche) in 1988, wild barley was harvested and the green ears were immediately threshed (Zohary in press). The observations in the French field confirmed those of Kislev which were done near Jerusalem. However, when the harvested barley was left to be sun-dried for 24 hours, only 2-3% of the sample had a broken fragment of the internode still attached to the disarticulation plane. Thus, Zohary has no difficulties in accepting the archaeobotanical evidence from Netiv Hagdud, Gilgal, and Jericho as reflecting the cultivation of two-rowed barley.

Finally, emmer wheat (*Triticum dicoccum*) is better represented in the charred remains from Tell Aswad (van Zeist and Bakker-Heeres 1985), the early phase, dated to 9,800-9,600 B.P. Since the wild form is missing altogether from the early phase, it is not surprising that in concluding a general survey of the early Neolithic plant husbandry, van Zeist was willing to speculate and suggest that:

> ...plant cultivation started inside the area in which wild cereals and pulses are found...the oldest archaeo-botanical evidence of domestic plants is from tell Aswad...It is clear that the farmers who settled at Aswad were already cultivating domestic emmer wheat. In addition...two rowed-hulled barley, field pea and lentil were grown...The presence of domestic emmer wheat implies that cultivation of this species (and probably other plant species) must have started already (long) before 7.800 B.C. One wonders whether the region covering SW Syria, northern Israel and Jordan...was an early, if not the earliest centre of domestication (van Zeist 1988: 56-58).

Interpretations

The overall geographic distribution of the early farming communities indicates that these were located along the modern boundary between the Mediterranean and the Irano-Turanian vegetational belts. We suggest to refer to this elongated stretch as the "Levantine Corridor," a term borrowed from paleontology where it encompasses the entire Levant. Paleobotanical, archaeozoological, geomorphic, and palynological evidence demonstrates that they lay within the limits of the Mediterranean belt during this time span. Thus it was in the "Levantine Corridor," from the Middle Euphrates through the Jordan Valley and into southern Jordan, that the first agricultural settlements were founded (Figure 6). On both sides, in the coastal range in the west and in the semiarid region in the east and south, small bands of hunter-gatherers carried on their way of life. A few of these sites have been excavated. One is the cave of Nacharini in the Anti-Lebanon mountains (Schroedeur 1977), the other Abu Madi I in southern Sinai (Bar-Yosef 1985). These

are small sites with a predominance of arrowheads (including Khiam points), perforators, some microliths, scrapers, and denticulates.

In conclusion, the Sultanian and the Mureybetian mark a departure from the Natufian way of life. Certain elements were preserved, such as round dwellings, the use of microliths, and the hunting of gazelle. But these are superficial resemblances. Different materials were used for building habitations and storage facilities are common in the later periods. The use of mudbricks resulted in rapid accumulation that led to decrease in tool densities per cubic meter. PPNA houses, whether of one or two rooms, stand as separate households. Several PPNA sites like Nahal Oren preserve the old way of a series of attached rooms as a compound. Such compounds are known from PPNB sites in southern Sinai and southern Jordan (Bar-Yosef 1984; Kirkbride 1978). They reflect the degree of kinship among the members of the group (Flannery 1972). Thus, it is clear that the real villages where an entire group (or "tribe"), a biological viable unit, lived together were located only in the strictly narrow part of the Levantine corridor (Bar-Yosef and Belfer-Cohen 1989; see Byrd this volume).

The question then is why change from the Late Natufian socioeconomic structure to the Sultanian way of life? If we accept the observation that population increase as expressed by site size (up to 2.5 or 3.0 hectars in the Sultanian) resulted from a fully sedentary way of life based on cultivation, then it is the onset of systematic cultivation instead of harvesting wild plants which may have triggered this process. Assuming that the Natufians had the knowledge of cereal and legume cultivation and perhaps even practiced it from time to time (Unger-Hamilton 1989), then it is not surprising that the stress period of the "Younger Dryas" forced them to increase production from smaller and less productive territories. As natural stands could not supply the needed surplus, it was probably the women of these groups (and the elders?) who initiated the planned cultivation. The best locations for such intentional cultivation were lake shores, which were watered every winter, and alluvial fans, with their high water table. This would explain the known location of all the large farming communities. The pluvial conditions of the early Holocene insured the success of annual farming and the relatively rapid population increase. The constant supply of cereals, one of the best weaning foods, ensured the survival of newborns. Within less than a 1,000 radiocarbon years the settlements became four times larger than their ancestral villages and the new subsistence strategy was carried across the Taurus and Zagros mountains into Anatolia and Iran.

Discussion: The Socioeconomic Reconstruction as Seen in 1991

Although many of the details are still missing, we have sufficient information to answer the "when" and "where" questions concerning the origins of agriculture. But we are far from providing a persuasive answer for the "why" question. There is no one simple model which can explain why this socioeconomic change finally happened. The answer in our view is in the form of a complex model based on a series of feedbacks between such factors as environmental change (caused by the rapid climatic fluctuations from 16,000 through 10,000 B.P.), the "island" situation of the Levant (which is like a "finger" of a wetter, narrow zone enclosed between the Mediterranean Sea and the deserts), the long-standing availability, predictability and accessibility of *r*-selected food resources such as the various cereals and legumes, and population pressure (caused by the specific late glacial prehistory of the region).

Gaps in our data sets require that we obtain better chronological control, especially for the Natufian period as well as additional archaeological materials which can be obtained by routine field archaeology. Examples for such projects would be the excavations of Late Natufian sites from across the Levantine Corridor, or the recovery of Natufian paleobotanical remains that are essential for determining whether cereals, legumes or other plants were grown intentionally or harvested in the wild. Large excavated surfaces at PPNA sites would provide needed information on household organization and social hierarchies. In the realm of experimental studies, we miss the quantitative information which can be obtained from growing the wild species of cereals, harvesting them in various times during the spring or fall, and employing different food-processing methods in the same way that flint knappers replicate alternative core reduction strategies.

In spite of these deficiencies we offer our current (Bar-Yosef and Belfer-Cohen 1991) explanation in the form of a combined model that takes into account the following factors:

(A) The changing environments which resulted from climatic shifts including (1) the increase in precipitation since 14,000 B.P., (2) a short crisis around 13,000 B.P., (3) a rapid increase in precipitation that culminated around 11,500 B.P., (4) the "Younger Dryas", a cold and dry period which lasted several centuries sometime between 11,000 and 10,000 B.P. and, (5) a rapid return to wetter conditions around 10,000 B.P. which ensured the existence of numerous small lakes and ponds and provided for the success of numerous PPNA sites along the Levantine Corridor. (B) The availability and predictability of many edible annuals (mainly cereals and legumes, which were abundant and have been exploited in the Levantine Corridor since at least 20,000 B.P.) and perennial plant resources (essentially fruits). (C) The existence of the necessary technology for collecting and gathering (baskets?) as well as for food processing. Grinding and pounding stone tools were used extensively from 20,000 B.P. onwards. (D) The behavior pattern of the common local game (e.g., the relative sedentism of the gazelle, fallow and roe deer). (E) "Demographic pressure" which did not necessarily mean a large increase in the number of people but was critical in that every ecozone of the Near East had been occupied during the Late Glacial (from ca. 14,500 B.P.). (F) The "multiplier effect" (Hole 1984) of the technological, social, and economic build-up since the Early Natufian, operating as a feedback mechanism over three thousand radiocarbon years. New inventions and innovations included the bow, arrows with aerodynamic tips, nets, a variety of hide working and basket fabricating bone tools (Campana 1989), the intensive use of mudbricks, polished and hafted celts, asphalt as an adhesive, etc., should be all taken into account as contributing to this process.

The information from these data sets can be interpreted in two different ways. The first approach, with which we do not concur, is gradualistic. It would deny any abruptness in the emergence of the new socioeconomic structure of Neolithic society. The holders of this view suggest that the technological changes observed from the Geometric Kebaran to the Natufian and the Early Neolithic represent a gradual, long-term, socio-economic process. Several scholars (e.g., Perrot 1968, 1983) claim that Natufian cultural attributes, such as the rounded houses, can also be seen in the Neolithic record. Moreover, the appearance of domesticates, both plants and animals, is seen as a series of small stages during which the Neolithic economy reached its full blown form, probably not before the end of the seventh millennium B.P.

The second approach, which we advocate here, views the emergence of the Natufian from a background of Epi-Palaeolithic hunter-gatherers as a revolutionary event which took place in a geographical, well-delineated Levantine "homeland." The rest of the Near East was also occupied by groups of hunter-gatherers at that time. We see the emergence of the Levantine PPNA as a rapid response within the southern part of the Levantine Corridor to changing conditions including the forcing effects of the "Younger Dryas" on the Late Natufian. In such a model of cultural "punctuated equilibrium," Natufian lifeways were established following a major socioeconomic crisis and continued in almost homeostasis until they went through another major crisis which forced the establishment of the earliest phase of the Neolithic. This phase is the "Khiamian." Eliminating this short phase by defining the assemblages as "mixed" would even exaggerate the observed differences in material culture between the Late Natufian and Early Neolithic. Such a comparison would justify the use of the term "Neolithic Revolution," even if its new meaning was removed from the original definition of V. Gordon Childe. This is a process which took place during two or three hundred radiocarbon years or less.

The above description would imply that we entirely favor "push" models (in which the explanation views humans reacting under stress conditions such as population growth, declining returns in collected vegetal foods, etc.). We also feel that there is sufficient evidence to suggest an interaction between "push" and "pull" models (Stark 1986). The latter perspective regards humans as investing in higher-yield species in return for their energy expenditure (when compared to local species). Thus, the cultivation of cereals can be interpreted by this model as intentional cultivation in ecozones where these species did not occur naturally.

Systematic cultivation on a year to year basis, under favorable climatic conditions, meant the creation of surplus. This could have been used on one hand for consumption by growing communities, and on the other hand as trade items (Runnels and van Andel 1988). The resulting long-distance exchange networks obtained rare commodities such as Anatolian obsidian. The growth of social power within the farming communities may have been an additional conse-quence along with the establishment of the ancestor cult, evidenced in the removal of skulls of old adults.

Finally, in our view, the identification in Levantine prehistory of a "core and periphery" relationship pertains to the question whether cereals and pulses were domesticated in one small area or across the Near East (see also Byrd, this volume). According to Zohary there was only one locus of domestication for each species (Zohary 1989). His observations, based on cytogenetic data, conform with our description of successive thresholds in the Natufian "homeland" as the decisive steps leading to the Neolithic Revolution. It is in the latter phases that domesticated cereals and legumes were moved across the Taurus into Anatolia and to the inter-montane valleys of the Zagros. In conclusion, the role of the direct ancestors of the Natufians, and their social and economic decisions taken around 13,000 B.P., proved to be crucial for the ensuing history of the entire Near East.

Acknowledgements

This paper is partially based on our previously published papers (Bar-Yosef and Belfer-Cohen 1989, Bar-Yosef and Belfer-Cohen 1991). We would like to thank T.D. Price and A.B. Gebauer for inviting us to participate in the session on "The Transition to Agriculture" held at the Annual Meeting of the Society for American Archaeology, April 24-28, 1991 in New Orleans. We are grateful to E. Hovers (Institute of Archaeology, Hebrew University) for her numerous useful comments on an earlier draft and to T.D. Price and Martha Tappen for editorial modifications.

References Cited

Albrecht, G.
   1988 An Upper Palaeolithic Sequence from Antalya in Southern Turkey. Results of the 1985 Cave Excavations in Karain B. In *L'Homme de Neandertal*, edited by J. Kozlowski, pp. 23–35. ERAUL, Université de Liege, Liege.
Arensburg, B., and Y. Rak
   1979 The Search for Early Man in Israel. *The Quaternary of Israel*, edited by A. Horowitz, pp. 201–209. Academic Press, New York.
Aufrey, J.C., E. Tchernov, and E. Nev
   1988 Origine du commensalisme de la souris domestique (*Mus musculus domesticus*) vis-à-vis de l'homme. *Comptes Rendus de l'Academie des Sciences Paris 307 (Série III)* 307, Série III:517–522.
Aurenche, O., J. Cauvin, F. Hours, and P. Sanlaville
   1981 Chronologie et organisation le l'espace dans le Proche Orient de 1200 a 5600 avant J.C. In *Préhistoire du Levant*, edited by J. Cauvin and P. Sanlaville, pp. 571–601. CNRS, Paris.
Baharav, D.
   1974 Notes on the Population Structure and Biomass of the Mountain Gazelle, *Gazella gazella gazella*. *Israel Journal of Zoology* 23:39–44.

1980 Habitat Utilization of the Dorcas Gazelle in a Desert Saline Area. *Journal of Arid Environments* 3:161–167.

1982 Desert Habitat Partitioning by the Dorcas Gazelle. *Journal of Arid Environments* 5:323–335.

1983 Reproductive Strategies in female Mountain and Dorcas Gazelles (*Gazella gazella gazella* and *Gazella dorcas*). *Journal of Zoology London* 200:445–453.

Baruch, U., and S. Bottema

1991 Palynological Evidence for Climatic Changes in the Levant ca. 17,000–9,000 B.P. In *The Natufian Culture in the Levant*, edited by O. Bar-Yosef and F.R. Valla, pp. 11–20. International Monographs in Prehistory, Ann Arbor.

Bar-Yosef, D.E.

1989 Late Paleolithic and Neolithic Marine Shells in the Southern Levant as Cultural Markers. In *Shell Bead Conference*, edited by C.F. Hayes, pp. 169–174. Rochester Museum and Science Center, Rochester, New York.

Bar-Yosef, O.

1975 The Epi-Palaeolithic in Palestine and Sinai. In *Problems in Prehistory: North East Africa and the Levant*, edited by F. Wendorf and A.E. Marks, pp. 363–378. SMU Press, Dallas.

1980 A Figurine From a Khiamain Site in the Lower Jordan Valley. *Paléorient* 6:193–200.

1981a The "Pre-Pottery Neolithic" Period in the Southern Levant. In *Préhistoire du Levant*, edited by J. Cauvin and P. Sanlaville, pp. 555–569. Editions CNRS, Paris.

1981b Epi-Palaeolithic Complexes in the Southern Levant. In *Préhistoire du Levant*, edited by J. Cauvin and P. Sanlaville, pp. 389–408. Editions CNRS, Paris.

1983 The Natufian of the Southern Levant. In *The Hilly Flanks and Beyond, Studies in Ancient Oriental Civilization*, edited by C.T. Young, P.E.L. Smith, and P. Mortensen, pp. 11–42. University of Chicago Press, Chicago.

1984 Seasonality Among Neolithic Hunter-gatherers in Southern Sinai. In *Animals and Archaeology: Herders and Their Flocks*, edited by J. Clutton-Brock and C. Grigson, pp.145–160. BAR International Series. BAR, Oxford.

1985 The Stone Age of the Sinai Peninsula. In *Studi di Paleontologia in Ocore i Salvatore M. Puglisi*, edited by M. Liverani, A. Palmieri and P. Peroni, pp. 107–122. Universita di Rome "La Sapienza," Rome.

1986 The Walls of Jericho: An Alternative Interpretation. *Current Anthropology* 27:157–162.

1987a Direct and Indirect Evidence for Hafting in the Epi-Palaeolithic and Neolithic of the Southern Levant. In *Le Main et l'outil, manche et emman elements préhistoriques*, edited by D. Stordeur, pp. 155–164. Maison de l'Orient, Lyon.

1987b The Late Pleistocene in the Levant. In *The Pleistocene Old World: Regional Perspectives*, edited by O. Soffer, pp. 219–236. Plenum Press, New York.

1989 The PPNA in the Levant — An Overview. *Paléorient* 15: 57-63.

1991 The Early Neolithic of the Levant: Recent Advances. *The Review of Archaeology* 12(2).

Bar-Yosef, O., and A. Belfer-Cohen

1988 The Early Upper Palaeolithic in the Levantine Caves. In *The Early Upper Palaeolithic in Europe and the Near East*, edited by J. Hoffecker and C. Wolf, pp. 23–41. BAR International Series 437. BAR, Oxford.

1989 The Origins of Sedentism and Farming Communities in the Levant. *Journal of World Prehistory* 3:447–498.

1991 From Sedentary Hunter-Gatherers to Territorial Farmers in the Levant. In *Between Bands and States*, edited by S.A. Gregg, pp. 181–202. Center for Archaeological Investigations, Carbondale.

Bar-Yosef, O., A. Gopher, M.E. Kislev, and E. Tchernov

in press. Netiv Hagdud—An Early Neolithic Village Site in the Jordan Valley. *Journal of Field Archaeology*.

Bar-Yosef, O., and N. Goren

1973 Natufian Remains in Hayonim Cave. *Paléorient* 1:49–68.

Bar-Yosef, O., and A. Killbrew

1984 Wadi-Sayakh—A Geometric Kebaran Site in Southern Sinai. *Paléorient* 10:95–102.

Bar-Yosef, O, and G. Martin

1979 La problème de la "Sortie des Grottes" au Natoufien: Répartition et localization des gisements épipaléolithiques du Levant Méditerranée. *Bulletin de la Société Préhistorique Française* 78:187–192.

Bar-Yosef, O., and E. Tchernov

1967 Archaeological finds and fossil faunas of the Natufian and Microlithic Industries at Hayonim Cave (Western Galilee, Israel). *Israel Journal of Zoology* 1:104–140.

Bar-Yosef, O., and F.R. Valla

1979 L'évolution du Natufien, Nouvelles Suggestions. *Paléorient* 5:145–152.

Bar-Yosef, O., and J.C. Vogel

1987 Relative and Absolute Chronology of the Epi-Palaeolithic in the Southern Levant. In *Chronologies in the Near East*, edited by O. Aurenche, J. Evin, and F. Hours, pp. 219–245. BAR International Series 379. BAR, Oxford.

Begin, Z.B., Y. Nathan, and A. Erlich

1980 Stratigraphy and Facies Distribution in the Lisan Formation: New Evidence from the South of the Dead Sea. *Israel Journal of Earth Sciences* 29:182–189.

Belfer-Cohen, A.

1988a The Appearance of Symbolic Expression in the Upper Pleistocene of the Levant as Compared to Western Europe. In *L'Homme de Neandertal. La Pensé,* edited by O. Bar-Yosef, pp. 25–29. ERAUL, Université de Liege, Liege.

1988b *The Natufian Settlement at Hayonim Cave.* Unpublished Ph.D. dissertaion. Hebrew University, Jerusalem.

1989 The Natufian Issue: A Suggestion. In *Investigations in South Levantine Prehistory,* edited by O. Bar-Yosef and B. Vandermeersch, pp. 297–307. BAR International Series 497. BAR, Oxford.

1991 Art Items from Layer B, Hayonim Cave: A Case Study of Art in a Natufian Context. In *The Natufian Culture in the Levant*, edited by O. Bar-Yosef and F.R. Valla, pp. 569–588. International Monographs in Prehistory, Ann Arbor.

Belfer-Cohen, A., A. Schepartz, and B. Arensburg

1991 New Biological Data for the Natufian Populations in Israel. In *The Natufian Culture in the Levant*, edited by O. Bar-Yosef and F.R. Valla, pp. 411–424. International Monographs in Prehistory, Ann Arbor.

Betts, A.

1982 A Natufian Site in the Black Desert, Eastern Jordan. *Paléorient* 8:79–82.

Betts, A., and S. Helms

1986 Rock Art in Eastern Jordan: "Kite" Carvings? *Paléorient* 12/1:67–72.

Blumler, M.A., and R. Byrne

1991 The Ecological Genetics of Domestication and the Origins of Agriculture. *Current Anthropology* 32:23–54.

Bottema, S.

1987 Chronology and Climatic Phases in the Near East from 16,000 to 10,000 BP. In *Chronologies in the Near East*, edited by O. Aurenche, J. Evin and F. Hours, pp. 295–310. BAR International Series 379. BAR, Oxford.

Bottema, S., and W. van Zeist

1981 Palynological Evidence for the Climatic History of the Near East 50,000–6000 BP. In *Préhistoire du Levant*, edited by J. Cauvin and P. Sanlaville, pp. 111–132 Editions CNRS, Paris.

Braidwood, R.G.

1975 *Prehistoric Men.* 8 ed. Scott, Foresmen, Chicago, Illinois.

Braidwood, R.J. , and L.S. Braidwood

1953 The Earliest Village Communities of Southwest Asia. *Journal of World History* 1(2):278–310.

Byrd, B.

1989 The Natufian: Settlement Variability and Economic Adaptations in the Levant at the End of the Pleistocene. *Journal of World Prehistory* 3:159–198.

Byrd, B.F., and A.N. Garrard

1990 The Last Glacial Maximum in the Jordanian Desert. In *Low Latitudes*, edited by C. Gamble and O. Soffer, pp. 78–92. The World at 18,000 BP, vol. 2. Unwin Hyman, London.

Campana, D.V.
  1989 *Natufian and Protoneolithic Bone Tools.* BAR International Series 494. BAR, Oxford.
Cauvin, J.
  1972 *Les Religions Néolithiques de Syro-Palestine.* Maisonn Euve, Paris.
  1977 Les fouilles de Mureybet (1971–1974) et leur signification pour les origines de la sédentarisation au Proche-Orient. *Annual of the American Schools of Oriental Research* 44:19–48.
  1978 *Les Prémiers villages de Syrie-Palestine du IXe au VIIIe millénaires av.* Maison de l'Orient, Lyon.
  1985 La question du "Matriarcat Préhistorique" et le role de la femme dans la préhistoire. In *La Femme dans le Monde Mediterraneen*, edited by A.M. V´érilhac, pp. 1-18, vol. 1 . Travaux de la Maison de l'Orient, Lyon.
  1989 La Neolithisation au Levant et sa Prémiere Diffusion. In *Neolithization*, edited by O. Aurenche and J. Cauvin, pp. 3 36. BAR International Series 516. BAR, Oxford.
  1990 Nomadisme Néolithique en Zone Aride: L'Oasis D'El Kowm (Syrie). *Resurrecting the Past*, edited by P. Matthiae, M. Van Loon, and H. Weiss, pp. 41–47. Nederlands Historisch-Archaeologisch Instituut, Istanbul.
Cauvin, M.-C.
  1974 Flèches à encodes de Syrie: Essai de Classification et d'Interprètation culturelle. *Paléorient* 2/2:311–322.
  1981 L'Epipaléolithique du Levant. Synthèse. In *Préhistoire du Levant*, edited by J. Cauvin and P. Sanlaville, pp. 439–441. CNRS, Paris.
Cauvin, M.-C., J. Coquegniot, and M.C. Nierle
  1982 Rapport preliminaire sur la campagne 1980 d'El Kowm 1. *Chiaers de l'Euphrate* 3:27–32.
Cauvin, M.-C., and D. Stordeur
  1978 *Les Outillages Lithiques et Osseux de Mureybet, Syrie.* Publications del'URA no. 17; Cahiers de l'Ephrate no. 1. CNRS, Paris.
Close, A.E.
  1978 The Identification of Style in Lithic Artefacts. *World Archaeology* 10:223–237.
Clutton-Brock, J.
  1979 The Mammalian Remains from Jericho Tell. *Proceedings of the Prehistoric Society* 45:135–157.
Crowfoot-Payne, J.
  1976 *The Terminology of the Aceramic Neolithic Period in the Levant. Terminology of Prehistory of the Near East.* Pre-print IX UISPP Congress, Nice.
  1983 The Flint Industries of Jericho. In *Excavations at Jericho*, edited by K.M. Kenyon and T.A. Holland, pp. 622–759. The British School of Archaeology in Jerusalem, London.
Davis, S., N. Goring-Morris, and A. Gopher
  1981 Sheep Bones from the Negev Epipaleolithic. *Paléorient* 8(1):87–93.
Davis, S.J.M.
  1982 Climatic Change and the Advent of Domestication of Ruminant Artiodactyls in the Late Pleistocene-Holocene Period in the Israel Region. *Paléorient* 8:5–16.
  1983 The Age Profile of Gazelles Predated by Ancient Man in Israel: Possible Evidence for a Shift from Seasonality to Sedentism in the Natufian. *Paléorient* 9:55–62.
Echegaray, G.J.
  1966 *Excavation en la Terraza de El-Khain (Jordania).* Part II. Madrid.
Edwards, P.C.
  1989 Revising the Broad Spectrum Revolution: And its Role in the Origins of Southwest Asian Food Production. *Antiquity* 63:225–46.
Edwards, P.C., S.J. Bourke, S.M. Colledge, J. Head, and P.G. Macumber
  1988 Late Pleistocene Prehistory in the Wadi al-Hammeh, Jordan Valley. In *The Prehistory of Jordan: The State of Research in 1986*, edited by A.N. Garrard and H.G. Gebel, pp. 525–565. BAR International Series 396. BAR, Oxford.
Flannery, K.V.
  1972 The Origins of the Village as a Settlement Type in Mesoamerica and the Near East: A Comparative Study. In *Man, Settlement and Urbanism*, edited by P.J. Ucko, R. Trigham, and G.W. Dimbleby, pp. 23–553, Gerald Duckworth & Co., London.

Garfinkel, Y., and D. Nadel
  1989 The Sultanian Flint Assemblage from Gesher and its Implications for Recognizing Early Neolithic Entities in the Levant. *Paléorient* 15/2:139–152.
Garrard, A., B. Byrd, and A. Betts
  1986 Prehistoric Environment and Settlement in the Azraq Basin: An Interim Report on the 1984 excavation season. *Levant* 18:1–20.
Garrard, A.N., A. Betts, B. Byrd, and C. Hunt
  1987 Prehistoric Environments and Settlement in the Azraq Basin: An Interim Report on the 1985 Excavation Season. *Levant* 19:5–25.
Goldberg, P.
  1986 Late Quarternary Environmental History of the Southern Levant. *Geoarchaeology* 1:225–244.
Goodfriend, G.A.
  1991 Holocene Trends in ^{18}O in Land Snail Shells from the Negev Desert and Their Implications for Changes in Rainfall Source Areas. *Quaternary Research* 35:417–426.
Gopher, A.
  1989 Neolithic Arrowheads in the Levant: Results and Implications of a Seriation Analysis. *Paléorient* 15/1:57–64.
Goring-Morris, A.N.
  1987 *At the Edge: Terminal Pleistocene Hunter-gatherers in the Negev and Sinai.* BAR International Series 361. BAR, Oxford.
  1991 The Harifian in the Southern Levant. In *The Natufian Culture in the Levant,* edited by O. Bar-Yosef and F.R. Valla, pp. 173–216. International Monographs in Prehistory, Ann Arbor.
Harrison, D.L.
  1968 *Mammals of Arabia.* 2 vols. Ernest Benn, London.
Helmer, D.
  1989 Le developément de la domestication au Proche-Orient de 9500 à 7500 B.P.: Les Nouvelles données d'El Kown et de Ras Shamra. *Paléorient* 15/1:111–121.
Henry, D.O.
  1974 The Utilization of the Microburin Technique in the Levant. *Paléorient* 2:389–398.
  1976 Rosh Zin: A Natufian Settlement near Ein Avdat. In *Prehistory and Paleoenvironents in the Central Negev,* edited by A.E. Marks, pp. 317–347. SMU Press, Dallas.
  1982 The Prehistory of Southern Jordan and Relationships with the Levant. *Journal of Field Archaeology* 9:417–444.
  1983 Adaptive Evolution within the Epipaleolithic of the Near East. In *Advances in World Archaeology,* edited by F. Wendorf and A.E. Close, pp. 99–160. Academic Press, New York.
  1985 Preagricultural Sedentism: The Natufian example. *Prehistoric Hunter-Gatherers: The Emergence of Complex Societies,* edited by T.D. Price and J.A. Brown, pp. 365–384. Academic Press, New York.
  1989 *From Foraging to Agriculture: The Levant at the End of the Ice Age.* University of Pennsylvania Press, Philadelphia.
Hillman, G.C., S. Colledge, and D.R. Harris
  1989 Plant Food Economy During the Epi-Palaeolithic Period at Tell Abu Hureyra, Syria: Dietary Diversity, Seasonality and Modes of Exploitation. In *Foraging and Farming: The Evolution of Plant Exploitation,* edited by G.C. Hillman and D.R. Harris, pp. 240–266. Hyman Unwin, London.
Hillman, G.C., and M.E. Davies
  1990a Domestication Rates in Wild-type Wheats and Barley Under Primitive Cultivation. *Biological Journal of the Linnaean Society* 39:39–78.
  1990b Measured Domestication Rates in Wild Wheats and Barley Under Primitive Cultivation and their Archaeological Implications. *Journal of World Prehistory* 4:157–222.
Hillman, G.C., E. Madayska, and J.G. Hather
  1989 Wild Plant Foods and Diet at Late Palaeolithic Wadi Kubbaniya: Evidence from the Charred Remains. In *Palaeoeconomy, Environment and Stratigraphy,* edited by F. Wendorf and R. Schilds, pp. 162–242. The Prehistory of Wadi Kubbaniya, vol. 2. SMU Press, Dallas.
Hole, F.
  1984 A Reassessment of the Neolithic Revolution. *Paléorient* 10:49–60.

Hopf, M.
   1983 *Excavations at Jericho.* The British School of Archaeology in Jerusalem, London.
Hours, F.
   1976 L'Epi-Paléolithique au Liban. Resultats acquis en 1975. In *Second Symposium on Terminology of the Near East,* edited by F. Wendorf, pp. 106–130. Congress for Pre- and Proto-Historic Sciences, Nice.
Kaufman, D.
   1986 A Reconsideration of Adaptive change in Levantine Epi-Paleolithic. In *The End of the Paleolithic in the Old World,* edited by L.G. Straus, pp. 117–128. BAR International Series 284. BAR, Oxford.
Kenyon, K.
   1957 *Digging Up Jericho.* Benn, London.
   1981 *The Architecture and Stratigraphy of the Tell.* Excavations at Jericho, vol. III. British School of Archaeology in Jerusalem, London.
Kenyon, K.M., and T.A. Holland
   1983 *The Pottery Phases of the Tell and Other Finds.* Excavations at Jericho, vol. V. British School of Archaeology in Jerusalem, London.
Kirkbride, D.
   1978 The Neolithic in Wadi Rumm: Ain Abu Nekheileh. In *Archaeology in the Levant,* edited by P.R.S. Moorey, and P.J. Parr, pp. 1-10. Warminster, England.
Kislev, M.E.
   1989 Pre-Domesticated Cereals in the Pre-Pottery Neolithic A Period. In *People and Culture Change,* edited by I. Hershkovitz, pp. 147–152. BAR International Series 508 (i). BAR, Oxford.
   in press. The Agricultural Situation in the Middle East in the 8th Millennium B.C. In *Exploitation des Plantes en Préhistoire,* edited by P.C. Anderson-Gerfaud. CRA, Valbonne.
Kislev, M.E., D. Nadel, and I. Carmi
   in press. Grain and Fruit subsistence at Early Epi-Palaeolithic Ohalo II, Jordan Valley. *Review of Palaeobotany and Palynology* 59.
Koucky, F.L., and R.H. Smith
   1986 Lake Beisan and Prehistoric Settlement of the Northern Jordan Valley. *Paléorient* 12:27–36.
Kuijt, I., J. Mabry, and G. Palumbo
   1991 Early Neolithic Use of Upland Areas of Wadi El-Yabis: Preliminary Evidence from the Excavations of Iraq ed-Dubb, Jordan. *Paléorient* 17 (1): 1-10.
LeChevallier, M., and A. Ronen
   1985 *Le site Natoufien-Khiamienne de Hatoula.* Les Cahiers de Centre de Recherche Français de Jerusalem 1.
Legge, A.J., and P.A. Rowley-Conwy
   1987 Gazelle Killing in Stone Age Syria. *Scientific American* 257(8): 88-95.
Leroi-Gourhan, A., and F. Darmon
   1987 Analyses palynologiques de sites archéologiques du Pleistocène final dans la vallée du Jourdain. *Israel Journal of Earth Science* 36:65–72.
Lieberman, D.R., T.W. Deacon, and R.H. Meadow
   1990 Computer Image Enhancement and Analysis of Cementum Increments as Applied to Teeth of *Gazella gazella. Journal of Archaeological Science* 17:519–533.
Margaritz, M., and G.A. Goodfriend
   1987 Movement of the Desert Boundary in the Levant from the Latest Pleistocene to the Early Holocene. In *Abrupt Climatic Change,* edited by W.H. Berger and L.D. Labeyrie, pp. 173–183. Reidel, Dordrecht.
Marks, A.E.
   1977 The Epipaleolithic of the Central Negev: Current Status. *Eretz Israel* 13:216–228.
Marks, A.E., and P. Larson
   1977 Test Excavations at the Natufian Site of Rosh Horsha. *Prehistory and Paleoenvironments in the Central Negev, Israel, Vol. II, The Advat/Agev Area,* edited by A.E. Marks, pp. 191–232. SMU Press, Dallas.

Marks, A.E., and A. Simmons

1977   The Negev Kebaran of the Har Harif. In *Prehistory and Paleoenvironments in the Central Negev, Israel, Vol II, The Adat/ Agev Area, Part 2*, edited by A.E. Marks, pp. 232–269. SMU Press, Dallas.

McCorriston, J., and F. Hole

1991   The Ecology of Seasonal Stress and the Origins of Agriculture in the Near East. *American Anthropologist* 93:46–94.

Meshel, Z.

1974   New Data about the "Desert Kites". *Tel Aviv* 1:129–143.

Mienis, H.K.

1977   Marine Molluscs from the Epipaleolithic and Harifian of the Har Harif. In *Prehistory and Paleoenvironments in the Central Negev, Israel, Vol II. The Advat/Agev Area, Part 2*, edited by A.E. Marks, pp. 347–354. SMU Press, Dallas.

1987   *Mollusks from the Excavation of Mallaha (Eynan). La faune du gisement natoufien de Mallaha (Eynan), Israel.* Association Paléorient, Paris.

Moore, A.

1985   The Development in Neolithic Societies in the Near East. *Advances in World Archaeology* 3:1–69.

Muheisen, M.

1988   Le Gisement de Kharaneh IV, note sommaire sur la Phase D. *Paléorient* 1:270–282.

Muheisin, M.

1985   L'Épipaleolithique dans le gisement de Khoraneh IV. *Paléorient* 11:149–160.

Nadel, D., and I. Hershkovitz

in press. Ohalo II—A Water Logged Early Epipalaeolithic Site in the Jordan Valley, Israel. *Current Anthropology* .

Nesteroff, W.D., D. Vergnaud-Grazzini, L. Blanc-Vernet, Ph. Olive, J. Rivault-Znaidi, and M. Rossignol-Strick

1983   Évolution climatique de la Méditerranée Orientale au cours de la dernier glaciation. In *Paleoclimatic Research and Models*, edited by A. Ghazi, pp. 81–97. D. Reidel, Boston.

Noy, T., A.J. Legge, and E.S. Higgs

1973   Recent Excavations at Nahal Oren, Israel. *Proceedings of the Prehistoric Society* 39:75–99.

Noy, T., J. Schuldrenrein, and E. Tchernov

1980   Gilgal I: A Pre-Pottery Neolithic A Ssite in the Lower Jordan Valley. *Israel Exploration Journal* 30:63–82.

Olszewski, D.I.

1986   *The North Syrian Late Epi-Palaeolithic: The Earliest Occupation at Tell Abu-Hureyra in the Context of the Levantine Epi-palaeolithic.* BAR International Series 309. BAR, Oxford.

1989   Tool Blank Selection, Debitage and Cores from Abu Hureya 1, Northern Syria. *Paléorient* 15:29–37.

Perevolotsky, Y., and D. Bahara

1987   The Abundance of "Desert Kites" in Eastern Sinai. An Ecological Analysis. In *Sinai Part 2: Human Geography*, edited by G. Gvirtzman, A. Shmueli, Y. Gradus, I. Beit-Arieh, and M. Har-El. Eretz and Tel Aviv University, Tel Aviv.

Perrot, J.

1966   Le gisement natoufien de Mallaha (Eynan), Israël. *L'Anthropologie* 70:437–484.

1968   La Préhistoire palestinienne. *Supplément au Dictionnaire de la Bible 8.* pp. 286–446. Letouzey and Ané, Paris.

1983   Terminologie et cadre de la préhistoire récente de Palestine. In *The Hilly Flanks and Beyond: Essays on the Prehistory of Southwest Asia*, edited by T.C. Young, P.E.L. Smith and P. Mortensen, pp. 113–122. Oriental Institute of the University of Chicago, Chicago.

1988   Les sépultures. In *Les Hommes de Mallaha (Eynan) Israël*, edited by J. Perrot, pp. 1-106. Association Paléorient, Paris.

Phillips, J.L.

1973   Two Final Paleolithic Sites in the Nile Valley and Their External Connections. *Papers of Geological Survey of Eygpt* 57.

Phillips, J.L., and E. Minz
   1977 The Mushabian. *Prehistoric Investigations in Gebel Meghara, Northern Sinai*, edited by O. Bar-Yosef and J.L. Phillips, pp. 149–183. Monographs of the Institute of Archaeology. Hebrew University, Jerusalem.

Rafferty, G.E.
   1985 The Archaeological Record on Sedentariness: Recognition, Development and Implications. *Advances in Archaeological Method and Theory.* 5:113–156.

Runnels, C., and T.H. van Andel
   1988 Trade and the Origins of Agriculture in the Eastern Mediterranean. *Journal of Mediterranean Archaeology* 1:83–109.

Saxon, E.C., G. Martin, and O. Bar-Yosef
   1978 Nahal Hadera V: An Open-air Site on the Israeli Littoral. *Paléorient* 4:253–266.

Schroedeur, H.B.
   1977 Nacharini, a Stratified Post-Natufian Camp in the Anti-Lebanon Mountains. Paper presented at the Society for American Archaeology, Chicago.

Shmida, A., M. Evenari, and I. Noy-Meir
   1986 *Hot desert ecosystems. Hot Desserts and Arid Shrublands*, edited by M. Evenari, pp. 379–387. Elsevier Science Publishers, Amsterdam.

Simmons, A.H., and G. Ilany
   1975–1977 What Mean These Bones? *Paléorient* 3:269–274.

Stark, B.
   1986 Origins of Food Production in the New World. In *American Archaeology Past and Future*, edited by D.J. Melzer, D.D. Fowler, and J.A. Sabloff, pp. 277–321. Smithsonian Institution Press, Washington, D.C.

Stekelis, M., and T. Yizrael
   1963 Excavations at Nahal Oren (preliminary report). *Israel Exploration Journal* 13:1–12.

Stordeur, D.
   1981 La contribution de l'industrie de l'os à la délimination des aires culturelles: L'example du Natoufien. In *Préhistoire du Levant*, edited by J. Cauvin and P. Sanlaville, pp. 433–437. CNRS, Paris.
   1988 *Outils et armés en os du gisement natoufien de Mallaha (Eynan), Israël.* Mémoires et Travaux du Centre de Recherche Française de Jérusalem, No. 6.
   1991 Le Natoufien et son évolution à travers; les artefacts en Os. In *The Natufian Culture in the Levant*, edited by O. Bar-Yosef and F.R. Valla, pp. 483–520. International Monographs in Prehistory, Ann Arbor.

Tangri, D., and G. Wyncoll
   1989 Of Mice and Men: Is the Presence of Commensal Animals in Archaeological Sites a Positive Correlate of Sedentism? *Paléorient* 15:85–94.

Tchernov, E.
   1984 Commensal Animals and Human Sedentism in the Middle East. *Animals and Archaeology*, edited by J. Clutton-Brock and C. Grigson, pp. 91–105. BAR International Series 202. BAR, Oxford.
   1991 Biological Evidence for Human Sedentism in Southwest Asia during the Natufian. In *The Natufian Culture in the Levant*, edited by O. Bar-Yosef and F.R. Valla, pp. 315–340. International Monographs in Prehistory, Ann Arbor.

Uerpmann, H.P.
   1981 Faunal Remains from Shams ed-din Tannira, a Halafian Site in Northern Syria. *Berytus* 30:3–52.
   1987 *The Ancient Distribution of Ungulate Mammals in the Middle East.* Dr. Ludwig Reichert Verlag, Weisbaden.

Unger-Hamilton, R.
   1989 The Epi-Palaeolithic Southern Levant and the Origins of Cultivation. *Current Anthropology* 31:88–103.

Valla, F.
   1981 Les establissements natoufiens dans le nord d'Israel. In *Préhistoire du Levant*, edited by Cauvin, J. and P. Sanlaville, pp. 409–420. CNRS, Paris.

1984 *Les Industries de Silex de Mallaha (Eynan) et du Natoufien dans le Levant.* Mémoires et Travaux du Centre de Recherche Français de Jérusalem 3. Association Paléorient, Paris.

1987 Chronologie absolue ét chronologies relatives dans le Natoufien. In *Chronologies du Proche-Orient*, edited by J. Evin, F. Hours, and O. Aurenche, pp. 267–294. BAR International Series S379i. BAR, Oxford.

Valla, F.R.

1988 Aspects du sol de l'abri 131 de Mallaha (Eynan). *Paléorient* 14/2:283–296.

Valla, F.R., H. Plisson, and R. Buxo i Capdevila.

1989 Notes Preliminaires sur les fouilles en cours sur la terrasse d'Hayonim. *Paléorient* 15/1:245–258.

van Zeist, W.

1988 Some Aspects of Early Neolithic Plant Husbandry in the Near East. *Anatolica* 15:49–68.

van Zeist, W., and J.A.H. Bakker-Heeres

1979 Some Economic and Ecological Aspects of the Plant Husbandry of Tell Aswad. *Paléorient* 5:161–169.

1985 Archaeobotanical Studies in the Levant: Neolithic Sites in the Damascus Basin, Aswad, Ghoraife, Ramad. *Paleohistoria (1982)* 24:165–256.

1986 Archaeobotanical Studies in the Levant. III. Late Paleolithic Mureybet. *Palaeohistoria* 26:171–199.

van Zeist, W., and S. Bottema

1982 Vegetational History of the Eastern Mediterranean and the Near East During the Last 20.000 Years. In *Palaeoclimates,Palaeoenvironments and Human Communities in the Eastern Mediterranean Region in Later Prehistory*, edited by J.L. Bintliff and W. van Zeist, pp. 231–277. BAR International Series 133. BAR, Oxford.

Watkins, T., D. Baird, and A. Betts

1989 Qermez Dere and the Early Aceramic Neolithic in N. Iraq. *Paléorient* 15/1:19–24.

Wigley, T.M.L., and G. Farmé

1982 Climate of the Eastern Mediterranean and Near East. In *Palaeoclimates, Palaeoenvironments and Human Communities in the Eastern Mediterranean Region in Later Prehistory*, edited by J.L. Bintliffe and W. Van Zeist, pp. 3–37. BAR International Series 133. BAR, Oxford.

Wright, H.E.

1977 Environmental Change and the Origins of Agriculture in the Old and New Worlds. In *The Origins of Agriculture*, edited by C.A. Reed, pp. 281–318. Mouton, The Hague.

Zarins, J.

1990 Early Pastoral Nomadism and the Settlement of Lower Mesopotamia. *Bulletin of the American School of Oriental Research* 31–65.

Zohary, D.

1969 The Progenitors of Wheat and Barley in Relation to Domestication and Agricultural Dispersal in the Old World. In *The Domestication and Exploitation of Plants and Animals*, edited by P.J. Ucko and G.W. Dimbleby, pp. 47–66. Gerald Duckworth & Co., London.

1989 Domestication of the Southwest Asian Crop Assemblage of Cereals, Pulses and Flax: the Evidence from the Living Plants. In *Foraging and Farming: The Evolution of Plant Domestication*, edited by D.R. Harris and G. Hillman, pp. 359–373. Unwin and Hyman, London.

in press. Domestication of the Neolithic Near East Crop Assemblage. In *Exploitation des Plantes et Préhistoire: Documents et Techniques*, edited by P. Anderson-Gerfaud. CNRS, Valbonne.

Zohary, D., and M. Hopf

1988 *Domestication of Plants in the Old World.* Oxford University Press, Oxford.

4

The Dispersal of Food Production Across the Levant

Brian F. Byrd

University of Wisconsin-Madison

The spread of agriculture outside of Southwest Asia has been a topic of extensive research and discussion. This is particularly true with respect to its dispersal to the northwest into central and northern Europe and westward in the circum-Mediterranean. Archaeologists working in these areas have presented relatively refined models which variously emphasize agricultural expansion or colonization, the adoption of agricultural products by indigenous hunter-gatherers, and the interplay of the two (e.g., Bogucki 1988; Lewthwaite 1986; Zvelebil and Rowley-Conwy 1986). Moreover, there exists a strong and healthy recognition of the complexity of the dispersal process and the presence of distinct regional variability (e.g., Ammerman and Cavalli-Sforza 1984; Bogucki 1987; Zvelebil 1989).

With a few notable exceptions (e.g., Bar-Yosef and Belfer-Cohen 1991, this volume; Hole 1984, 1987; Kislev 1984; Voigt 1983), however, there has been limited discussion of the spread of agricultural into a diverse range of ecological niches within Southwest Asia. This is potentially a fruitful topic of research, particularly with respect to the Levant, where archaeological field research has been extensive. Within the Levant, an area of more than 100,000 sq km, it took more than two millennia for domestic crops and herd animals to become a major component of the economy (e.g., Moore 1985, Bar-Yosef and Belfer-Cohen 1989a, 1991; Hole 1984).

Several questions remain unanswered with respect to the dispersal processes within the Levant. For example, did local indigenous hunter-gatherers adopt cultigens throughout the fertile areas of the Levant or did agricultural populations expand out from discrete centers of domestication, colonizing new territories? And what role did the subsequent adoption of herd animals play in the spread and entrenchment of food producing villages? Of particular interest for this discussion are questions concerning differences between the fertile and arid portions of the Levant.[1] This consideration entails an understanding of the nature of interaction between agricultural villages in the fertile areas and the exploitation of the arid margins, and how the late introduction of herding in the latter area further altered the existing subsistence and settlement strategies.

This paper briefly summarizes the current knowledge regarding these issues and offers some possible, albeit speculative, hypotheses. This will be done by first exploring how agriculture varied and may have spread within the fertile areas of the Levant during the Early Neolithic (ca. 10,200 b.p. to 8000/7,500 b.p.). Then the dispersal of food production into the arid margins of the Levant and the nature of interaction with settled agriculturalists in the fertile area will be considered.

The Fertile Levant

In Southwest Asia, research by botanists and plant geneticists has identified the wild progenitors of most of the initial or "founder" domestic crops, along with their geographical distribution and ecological range (e.g., Ladizinsky 1989; Zohary 1989; Zohary and Hopf 1988). The initial crops of the early Neolithic include three cereals (emmer wheat, einkorn wheat, barley), five pulses (lentil, broad bean, chickpea, field pea, bitter vetch), and flax. Of this crop complex, only the wild progenitor of chickpea does not occur within the Levant (Ladizinsky and Adler 1976). Moreover, four of these founder crops—barley, emmer wheat, lentil, and field pea—were consistently exploited within the Levant during the early Neolithic and can be considered the principle crops of the region.

Zohary (1989) contends that the domestication of each of these plants, with the possible exception of lentils (see Ladizinsky 1989; Pinkas et al. 1985), appears to have been a solitary, monophyletic event. This assertion is based on cytogenetic studies of the wild progenitors which reveal that they are more variable than the domestic crops, and that there is an absence of parallel evolution of key traits. If Zohary's single origin hypothesis is correct, then the mechanisms by which each of these founder crops spread from their initial locales of domestication within Southwest Asia is an exceedingly interesting subject of investigation. It also raises the question of whether each species was domesticated separately or if several were domesticated together in one locale.[2]

Pre-Pottery Neolithic A (PPNA)

The onset of *cultivation* appears to have occurred in the PPNA (radiocarbon dated between 10,200/10,300–9,600/9,300 b.p.) or just prior to it (e.g., Bar-Yosef and Belfer-Cohen 1989a, 1991, this volume; Kislev and Bar-Yosef 1988; Zohary and Hopf 1988). The validity of the morphological evidence for domestication is subject to differing interpretations, particularly with respect to barley (Kislev, 1989; Kislev et al. 1986; Miller 1991; Zohary and Hopf 1988). These cultivated products, however, represented only a modest contribution to an economy focused primarily on the intensive exploitation of a wide variety of local wild plants and animals (Bar-Yosef 1989; Bar-Yosef and Belfer-Cohen 1989a; van Zeist and Bakker-Heeres 1982).

Sedentary, food-producing villages appear restricted both in number and geographical distribution during the PPNA. To date, only a few PPNA village sites have been excavated (Gigal, Netiv Hagdud, Jericho, and Tell Aswad). They have yielded evidence for the systematic exploitation of the principle Levantine crops (Bar-Yosef and Kislev 1989; de Contenson et al. 1979; Hopf 1983; Noy 1989). The sites are all located in a relatively restricted portion of the south-central Levant (Figure 1). This area is currently considered the most viable candidate for a local center

of domestication (Bar-Yosef and Belfer Cohen 1991, this volume; McCorriston and Hole 1991; van Zeist 1988; Zohary and Hopf 1988). Along the Euphrates in the northern Levant, contemporary Mureybet (and the earlier occupation at Abu Hureyra) are the only other sites of comparable size, complexity, thickness of deposit, and archaeobotanical remains. These sites, however, have been interpreted as sedentary settlements lacking domesticates and apparently intensively exploiting wild einkorn wheat (Cauvin 1978; Hillman et al. 1989; van Zeist and Bakker-Heeres 1984a).[3]

The question thus arises as to how far agricultural products dispersed across the Levant and what were the mechanisms involved during the PPNA—a period of between 600 and 1000 years? Did the spread occur through colonization or by the adoption of domestic plants by local hunter-gatherer groups? Unfortunately, the present state of knowledge restricts this discussion to the realm of hypotheses. I would predict, however, that the initial spread of agriculture occurred quite slowly through contact and exchange with indigenous hunter-gatherers, primarily along the forest-steppe boundary of what Bar-Yosef and Belfer-Cohen (1989a:483-484) have termed the "Levantine Corridor" (Table 1). The hunter-gatherers most likely to initially adopt agricultural products would have been those in logistically organized settlements (Binford 1982) located within ecological niches well-suited for agriculture. Such locales no doubt included abundant surface water that could be utilized for small field agriculture (e.g., Sherratt 1980; van Zeist 1988). In these locales, where populations were situated for extended periods of each year and were intensively collecting local wild plants, scheduling problems related to integrating domestic crops would have been minimal. Hence, they would have been able to incorporate founder crops into existing subsistence systems.

Throughout much of the remainder of the Levant, particularly portions of the forested western edge and throughout the steppe and desert, relatively few excavations have been undertaken at PPNA sites and almost no plant remains have been recovered. Present evidence does suggest, however, that hunter-gatherers with more residentially organized economic systems appear to have continued to exploit relatively large foraging territories (Bar-Yosef and Belfer-Cohen 1989a). Whether these groups incorporated domestic plants into their economy on a very modest scale is currently unknown, but an intriguing possibility.

Pre-Pottery Neolithic B (PPNB)

The subsequent PPNB, which lasted between 1300 and 1600 years, witnessed the spread of sedentary agricultural villages throughout the area where rainfall

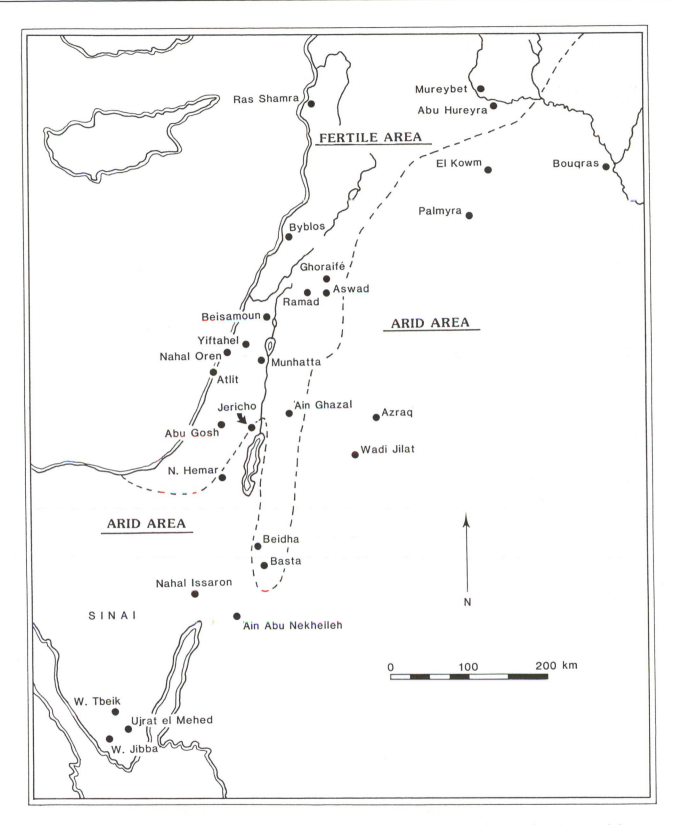

Figure 1. Map of the Levant showing prominent early Neolithic sites and a rough estimate of the boundary (dashed line) between the fertile area (the forest and the wet steppe) and the arid area (the dry steppe and desert).

Table 1. Summary sketch outlining the hypothetical dispersal of agriculture across the Levant. [1]

PPNA (ca. 10,200–9,600/9,300 b.p.)		
	Fertile Area	**Arid Area**
Predominant Settlement Type	Agricultural villages in S-C , H-G sites (usually seasonal) elsewhere	Short-term seasonal encampments and/or task specific sites
Mode of Spread	Contact & Exchange	If any, contact & exchange
Timing of Spread		
Plants	Slow	None probably
Herd Animals	None	None
Degree of Reliance		
Plants	Modest	None probably

PPNB (9,600/9,300 - 8,000/7,500 b.p.)		
	Fertile Area	**Arid East-central & Southern Areas**	**Arid Northeast Area**
Predominant Settlement Type	Agricultural villages replacing seasonal settlements over time	Seasonal encampments and various hunting sites [2]	A few Late and Final villages, seasonal and hunting sites earlier?
Mode of Spread	Contact & Exchange	Contact & Exchange, or trade?	Colonization
Timing of Spread			
Plants	From Early through Late	From Middle or Late	Late
Herd Animals	From Middle through Late	From Late/Final	Late
Degree of Reliance			
Plants	Extensive throughout, but varied emphasis	Very limited, primarily cereals	Extensive, primarily cereals
Herd Animals	Extensive, except in western extreme?	Limited (perhaps none in S)	Extensive

[1] Chronological subdivisions of the PPNB include Early (9,600–9,200 b.p.), Middle (9,200–8,500 b.p.), Late (8,500–8,000 b.p.) and Final (8,000–7,500 b.p.) (Cauvin 1987; Rollefson 1989).

[2] Hunting sites in the east-central Levant include game drive "kites," hunting blinds, and overlook locales.

agriculture was viable (e.g., Bar-Yosef and Belfer-Cohen 1989b; Moore 1985, 1989; Rollefson 1989). Overall reliance on domesticated plants steadily increased, the herding of ovicaprids was integrated into the economy, and clear regional patterns of exploitation emerged (Table 1).[4]

Where preservation permits, the four principle domesticated crops of the Levant (emmer wheat, barley, lentil, and field pea) are consistently well-represented at PPNB sites (Donaldson 1986; Hopf 1983; Nissen et al. 1987; Simmons et al. 1988; van Zeist 1986, 1988; van Zeist and Bakker-Heeres 1982, 1984a 1984b; van Zeist and Waterbolk-van Rooijen 1985). Einkorn wheat, bitter vetch, and chickpeas are typically rare or absent, although there are some notable exceptions such as Mureybet and Jericho where einkorn wheat is reported to occur in large numbers (Hopf 1983; van Zeist and Bakker-Heeres 1984a).[5] In addition, broad beans are well-represented only at Yiftahel (recovered in large numbers from a cache in a burned building) in the forested hills of the Western Levant (Garfinkle et al. 1987; Kislev 1985).

Despite the modest sample of sites, the relative frequency of archaeobotanical remains of particular species varies noticeably between settlements.[6] Barley predominates over emmer wheat in more semi-arid locales, due to productivity in less than optimal growing situations. Moreover, cereals are more common than pulses at most sites, particularly in the more semi-arid areas. At several sites (Ain Ghazal,

Yiftahel, and Ramad), however, legumes represent a significantly larger percentage of the archaeobotanical crop remains (Garfinkel et al. 1987; Simmons et al. 1988; van Zeist and Bakker-Heeres 1982). If this is indeed a reflection of the relative contribution of pulses versus cereals, then it may indicate that economic strategies varied considerably between local areas. Based on this small sample of sites, settlements in the well-forested areas (particularly the western edge of the Levant) may have had the greatest reliance on legumes.

The integration and increasing importance of domesticated ovicaprids in the economy takes place between 9,000 b.p. and 8,000 b.p. (Bar-Yosef and Belfer-Cohen 1989b; Rollefson 1989). This emphasis on ovicaprids occurs initially at a series of sites situated along the forest-steppe boundary of the central Levant (e.g., Helmer 1989; Kohler Rollefson 1989). The addition of this new source of protein may have offset the local depletion of wild game and possibly a decline in the productivity of pulse crops.[7]

Settlements in other areas (particularly the western third of the Levant and the arid margins), however, do not appear to integrate herd animals into the economy until somewhat later, during the late and final PPNB (e.g., Bar-Yosef 1984; Davis 1988; Garrard et al. 1988a; Horwitz 1987; Legge and Rowley-Conwy 1987).

Agricultural villages are situated in a diverse range of ecological settings throughout the fertile areas of the Levant by the end of the late PPNB, ca. 8000 B.P., over 2,200 years after the onset of the Neolithic. Both large and more moderate size settlements occur, the largest situated primarily in steppic locales, particularly in the eastern Levant. The timing of village dispersal is still poorly understood and debated, due in large part to the lack of chronological resolution for early PPNB occupation, particularly in the southern Levant (Cauvin 1987, 1989; Rollefson 1989).

The mechanisms by which agriculture spread throughout the fertile Levant during the PPNB are as yet undemonstrated. I would hypothesize that sedentary food-producing adaptations dispersed during the PPNB both by colonization and by local indigenous hunter-gatherers adopting domestic products and a food producing economy (Table 1). If agriculture continued to disperse through contact and exchange with indigenous hunter-gatherers as I have hypothesized for the PPNA, then it would have had to shift from a relatively residentially organized settlement system to a logistically organized one in order for food producers to become sedentary (Binford 1982).

No unequivocal examples of this shift are currently known, primarily because mustering strong archeological evidence is difficult and often open to multiple interpretations. The shift from round to square architecture, however, may be considered an indicator of changes in settlement mobility in the Levant. During the early Neolithic, the shift from round to rectangular architecture occurs considerably earlier at some sites in the fertile area and never in most portions of the arid margins (e.g., Bar-Yosef 1985; Garrard et al. 1986; Goring-Morris and Gopher 1983; Kirkbride 1978). For example, the middle to late PPNB site of Beidha witnessed a shift over time from architecture characterized by round structures with considerable wood construction to more substantial rectangular buildings of stone (Kirkbride 1966; Byrd, in prep). In contrast, contemporary large sedentary villages such as PPNB Jericho and Ain Ghazal are characterized by a rectangular architecture (Byrd and Banning 1988). Although it is risky to try and correlate round structures with less permanent settlement, could the initial building phase at Beidha reflect the architectural tradition of local hunter-gatherers shifting to a more sedentary lifestyle? Of course other lines of evidence, beyond the scope of this paper, are needed to rigorously address this issue.

Colonization by extant agriculturalists also may have been a major mechanism contributing to the spread of food production. Village population growth, which has been suggested by several scholars, may have spurred this process (e.g., Bar-Yosef and Belfer Cohen 1989b; Moore 1989; Smith et al. 1984). This growth could have included both the in-filling of niches between extant villages and the colonization of new portions of the fertile areas of the Levant. Colonization, regardless of its impetus, would have impinged on existing hunter-gatherer foraging territories, no doubt forcing them to more intensively exploit their remaining territory and focus considerable attention on the more marginal arid fringes of the region.

The Arid Margins of the Levant

Early Holocene adaptations in the dry steppe and desert of the Levant contrast markedly with settlement in the fertile areas. The more notable characteristics of arid land habitation include variability in occupation intensity over the millennium, the distinctiveness of local regional settlement and subsistence adaptation, and the delayed adoption of agricultural products (Table 1). The most extensive research has been conducted in the semi-arid Negev and Sinai; until recently much less was known about early Holocene occupation on the eastern Jordanian and Syrian plateau (Figure 1).

Pre-Pottery Neolithic A (PPNA)

Despite considerable survey and excavation (e.g., Akazawa 1979; Bar-Yosef 1985; Betts 1989; Garrard et al. 1988a; Henry 1982; Marks 1977; Simmons 1981), only limited evidence of occupation temporally coeval with the PPNA has been recovered in the arid regions of the Levant. In addition, the Harifian, a late Natufian adaptation restricted spatially to the Negev, may overlap slightly in time with the beginning of the PPNA (Bar-Yosef and Belfer-Cohen 1989a:475-476). There is no compelling reason, however, to consider that these areas were abandoned during this period. Rather it is probable that occupation was less intense and that settlements were more ephemeral. These regions no doubt witnessed either year-round occupation by small, mobile hunter-gatherer bands or seasonal/task-specific exploitation within a larger territory that included a portion of the fertile area.

Pre-Pottery Neolithic B (PPNB)

In contrast, evidence for PPNB occupation is abundant throughout much of the arid region and local variation in adaptation can be recognized. PPNB settlements are common in the Negev and Sinai, primarily during the latter half of the PPNB (Bar-Yosef 1981, 1985). Excavation data from these PPNB settlements is best known from the southern Sinai (Bar-Yosef 1984) and fieldwork in the central Negev (e.g., Goring-Morris and Gopher 1983; Servello 1976). In the southern Sinai, PPNB occupation is considered to represent a closed hunter-gatherer settlement system with winter/summer seasonal movements to exploit varied local resources (Bar-Yosef 1985:116). Although grinding stones are well represented at the summer encampments, the lack of preserved archaeobotanical remains has precluded exact determination of the resources exploited. Bar-Yosef (1984:158, 1985:115) has speculated that domestic grain may have been consumed and that this resource could have been obtained from village communities to the north in exchange for either marine shells or wild game meat. Faunal exploitation throughout the Negev and Sinai is focused on the hunting of wild game and there is no evidence of domestic animals being introduced into the region during the PPNB (Tchernov and Bar-Yosef 1982; Dayan et al. 1986).

The research results from several projects along the eastern arid margin of the Levant—the Jordanian and Syrian plateau—are providing further insights into the varied patterns of exploitation in the dry steppe and desert. Along the northern edge of the plateau (from the El Kowm Basin northward) in areas where rainfall averages less than 125 mm per annum, both surface scatters and substantial PPNB tell settlements occur (Cauvin et al. 1979). Extensive excavations have been undertaken at two PPNB tells in the El Kowm Basin (Dorneman 1986; Stordeur 1989) and at the large PPNB village site of Bouqras along the Euphrates

(Akkermans et al. 1981). These late and final PPNB settlements (post 8,500 B.P.) are sedentary farming villages characterized by elaborate rectangular architecture and the presence of domestic ovicaprids, wheat, and barley (Stordeur 1989; van Zeist 1986; van Zeist and Waterbolk-van Rooijen 1985). Although precipitation was greater during the early Holocene, it still may not have been sufficient for dry farming in these areas. Therefore it is possible that locally available surface water was utilized (van Zeist 1986; van Zeist and Waterbolk-van Rooijen 1985).

The southeastern portion of the Levant (south of the Azraq Basin) has witnessed only limited field research. Investigations in the east-central Levant (from the Azraq Basin northward to the Palmyra Basin), however, are revealing extensive evidence of PPNB occupation (Figure 1). Two long-term field projects in the Azraq Basin have provided considerable evidence of PPNB occupation in the arid eastern margins— Alison Betts' research in the northern portion of the basin and Andrew Garrard's fieldwork in the central and western areas (e.g., Betts 1989; Garrard et al. 1988a).[8] The latter project is providing the most intriguing information on changes in early Holocene arid subsistence patterns (Garrard et al., 1985, 1986, 1987, 1988a, 1988b).

In the east-central Levant, PPNB sites include semi-permanent, seasonal settlements, hunting camps, and associated blinds and game drive "desert kites" (e.g., Betts 1989; Garrard et al. 1988a; Helms and Betts 1987). Hunting, particularly of gazelle, is a major economic activity. Of extreme interest is the fact that domestic cereals, particularly barley, have been recovered at nearly all sites dating from the middle PPNB into the late Neolithic (9,500 b.p.–7,000 b.p.) excavated by Garrard in the western and central Azraq Basin (Garrard et al. 1988b:45-46). Two plausible explanations have been suggested for the appearance of these cereal crops outside their modern natural distribution (Garrard et al. 1988a:334; see also Bar-Yosef 1984). First, they were being grown locally, but this would be possible only if conditions were considerably moister than today or if limited surface water exploitation was being undertaken. Alternatively, the cereals may have been grown further to the west outside the Azraq Basin in areas more suitable for rainfall agriculture. This would have required the inhabitants of the Jilat area either to be exploiting both zones as part of a yearly settlement cycle, or for them to have acquired the cereals through trade with sedentary farming communities in the west.

The integration of moderate quantities of domestic cereals into an essentially hunter-gatherer economy appears to continue in the east-central arid margins throughout the PPNB and into the next millennium. In addition, at the end of the PPNB another change in subsistence occurs with the introduction of non-indigenous ovicaprids. At more than one site dated between 8200 B.P.–7800 B.P., significant quantities of ovi-caprid remains have been recovered. Based on their occurrence well outside their native habitat, Garrard has suggested that these are domestic animals (Garrard et al. 1988a:332-333). This exploitation of domestic ovi-caprids occurs well over 500 years later than at nearby sedentary villages such as Ain Ghazal (e.g., Helmer 1989; Kohler Rollefson 1989). The herding of these animals does not, however, replace wild game as a food resource but rather appears to supplement them. In fact, there is no evidence of a dramatic decline in gazelle due to overkill in game drives during this period (Garrard et al. 1988a), as has been previously suggested (Legge and Rowley-Conwy 1987).

So what were the mechanisms by which agriculture spread into the arid dry steppe and desert of the Levant? I suggest that both colonization and the local adoption of domestic products contributed to this process. Given the present evidence, regional patterns are apparent (Table 1). A few sites, such as Bouqras along the Euphrates and El Kowm, in the northeast margins of the Levant, appear to have been colonized by sedentary agriculturalists. The origin of these populations is uncertain, but the excavators have noted parallels with villages further to the east outside of the Levant in northern Iraq (Akkermans et al. 1981:370; Stordeur 1989:431). The establishment of these communities, particularly Bouqras, may be a component of agricultural village expansion into the more arid portions of the Euphrates, concomitant with the use of irrigation systems.

In contrast, I suggest that the mechanisms for the introduction of domestic cereals and subsequently ovicaprids into the arid east-central Levant, and possibly the Negev and Sinai, was not colonization by farmers establishing new settlements.[9] Rather, long-term contact and interaction between local hunter-gatherers and sedentary farmers in the fertile areas was taking place. Although it cannot be demonstrated at present, I suspect that we are seeing the selective integration of domestic resources into the economy of indigenous hunter-gatherers, primarily to diversify the diet. The addition of new resources with a more predictable yearly return rate may have been considered as a means of reducing risk. This is particularly so, given the arid setting and the wide fluctuations in the amount and location of annual rainfall. Cereals may have been a particularly attractive addition since they are high in carbohydrates, a relatively scarce resource in these arid areas (Speth 1983).

Outside of these areas, the only early Neolithic sites in the Levant with domestic plants and animals are sedentary village communities. The latter have fully enclosed rectangular domestic structures, plastered floors, subfloor burials, and a rich ideological tradition evidenced archaeologically by the removal of skulls after burial, the subsequent plastering of some of them, and the production of human statues and animal figurines. Settlements in the east-central Levant and the Negev and Sinai lack these attributes. Although it not impossible that colonists from sedentary agricultural communities moved into these areas during the PPNB, their reversion back to an essentially seasonally mobile hunter gatherer economy would have a number of interesting implications.

Why is there, however, such extensive evidence for exploitation of the dry steppe and desert during the PPNB and only limited evidence in the immediately preceding PPNA? If during the PPNB colonization was a significant factor in the spread of sedentary agricultural settlements within the fertile Levant, local hunter-gatherer populations would have been displaced. This could have resulted in higher hunter-gatherer population densities in arid areas such as the east-central Levant. In order to survive, they would have been forced to both intensify and diversify exploitation of these marginal areas. The appearance during the PPNB of numerous "desert kites" which were used for gazelle game drives is an example of such intensification. The small scale and slow adoption of domestic products can be considered an example of resource diversification. If domestic cereals were acquired by trade, then what was exchanged in return? Such trade could have included dried gazelle meat, obtainable in large quantities at desert-kite kill sites (Bar-Yosef 1984; Bar-Yosef and Belfer-Cohen 1989b). This pattern would fit with observations that the relative cost of obtaining protein versus carbohydrates were reversed for hunter-gatherers versus agriculturalist (e.g., Bogucki 1987; Speth 1983).

Pastoral nomadism has been suggested by Zarins (1989:43, 1990:53-54) as the stimulus for ninth millennium B.P. occupation of the arid eastern margins of the Levant. This assertion is questionable since there is extensive evidence of middle to late PPNB occupation in these areas—including game drives, hunting overlooks, and seasonal encampments—prior to the introduction of ovicaprids between 8,200 B.P. and 7,800 B.P. (e.g., Betts 1989; Helms and Betts 1987; Garrard et al. 1988a). Moreover, ovicaprids are initially integrated into the economy on a modest scale, and hunting continued to provide a major source of food. Arid land pastoralism appears to be a subsequent adaptation which no doubt had far-reaching implications for the interaction of settled villagers and remnant populations of hunter-gathers.

Conclusion

The ascendance of food-producing economies throughout the Levant was a long process entailing a series of stages that took over two millennia. The spread was no doubt complex and episodic, involving both the incorporation of domestic plants and animals into local hunter-gatherer economies and the founding of new settlements by agriculturalists. The questions that arise and the data to be examined regarding this dispersal in Southwest Asia share much in common with on-going research into the spread of agriculture outside the region.

It is only recently that archaeological data in the Levant have become adequate to address these issues. A high degree of interaction and exchange of information and possibly materials characterized the PPNB of the Levant (e.g., Bar-Yosef and Belfer-Cohen 1989b; Cauvin 1989; Rollefson 1989). Such dramatic differences in material culture between hunter-gatherers and farming communities distinguished in Keeley's study of the spread of agriculture in northern Europe (Keeley, this volume) are not apparent in the Levant.

As a result, integrated and detailed studies of variation in regional social organization, ideology, and settlement and subsistence are needed to elucidate the mechanisms behind the dispersal of agriculture within the Levant. The results of future research explicitly focused on this issue will certainly reveal patterns of interaction and adaptation not yet fully realized.

Regional studies which incorporate the climatic gradient from the forest and the dry areas along the eastern edge of the Levant will be of particular value. In addition, rigorous intrasite studies aimed at elucidating subtle changes in subsistence strategies and settlement mobility during the span of occupation are also needed. Such research will be strengthened if recent advances in the study of subsistence patterns are also employed. For example, phytolith and lipid residue analysis may provide fresh insights into plant resource exploitation, particularly in contexts where paleobotanical remains are poorly preserved.

Acknowledgements

The scholarly contributions of Andrew Garrard and Ofer Bar-Yosef, along with discussion that I have had with them, have made this paper possible. In addition, I am grateful for the comments on earlier drafts provided by William Belcher, Doug Price, and especially my wife Seetha Reddy.

Notes

[1] For the purposes of this paper, the fertile area is considered the forest and wet steppe, while the arid margins include the dry steppe and the desert. Of course, the distinction is somewhat arbitrary and the necessity of reconstructing the environment of the Neolithic adds further ambiguity.

[2] Even if multiple domestications of particular species did occur, the spread from these locales still would be of considerable interest. It would also be a much more complicated situation.

[3] The identification of intensive wild plant collection, cultivation, morphological domestication and archaeological evidence for them remains a compelling and much needed subject for research (e.g., Hillman and Davies 1990).

[4] The four phase PPNB chronological scheme proposed by Cauvin (1987) is utilized in this article. It consists of early (9,600–9,200 b.p.), middle (9,200–8,500 b.p.), late (8,500–8,000 b.p.), and final (8,000–7,500 b.p.) PPNB (see also Rollefson 1989).

[5] In contrast, these resources are often important domesticated products in Anatolia and the Zagros (e.g., Flannery 1969; van Zeist 1972, 1988).

[6] It can be, however, hazardous to make uncritical comparisons of relative frequencies of archaeobotanical remains between occupation horizons due to such factors as differences in food preparation techniques, preservation, and sample types. Each of these factors may cause plant species to be differentially preserved both within and between sites and this impedes attempts to make such inferences (e.g., Hastorf and Popper 1988) Yet, given the considerable ecological variability between micro-environments, distinct ecological constraints of the founder crops, and that different local wild resources were available to compliment the diet, regional patterns in resources emphasis are not unexpected.

[7] At all Neolithic sites with multiple phases of occupation, the relatively frequency of pulses versus cereals declines over time. There are number of possible of possible explanations for this trend (e.g., van Zeist and Bakker-Heeres 1982:237), including the possible overuse and depletion of local agricultural soils. Whatever the cause, this trend suggest that this source of protein may have become less important during the occupation span of a settlement.

[8] The Azraq Basin is a shallow depression covering 12,000 sq. km of north-central Jordan. It begins 30 km east of Amman and the large Neolithic site of Ain Ghazal and includes a climatic gradient running from dry steppe into desert.

[9] Of course, due to the lack of preservation, there is no evidence of domestic plants in the PPNB of the Negev and Sinai (Bar-Yosef 1985). Moreover, based on the sites excavated to date, domestic animals appear not to be introduced into this region until post PPNB (Tchernov and Bar-Yosef 1982).

References Cited

Akazawa, T.
  1979 Prehistoric Occurrences and Chronology in Palmyra Basin, Syria. In *Paleolithic Site of Douara Cave and Paleogeography of Palmyra Basin in Syria,* edited by K. Hanihara and T. Akazawa, pp. 201-220. Bulletin 16. University Musuem, University of Tokyo, Tokyo.
Akkermans, P.A., J.A.K. Boerma, A.T. Clason, S.G. Hill, E. Lohof, C. Meiklejohn, M. LeMier
  1981 Bouqras Revisited: Preliminary Report on a Project in Eastern Syria. *Proceedings of the Prehistoric Society* 49:335-372.
Ammerman, A.J. and L.L. Cavalli-Sforza
  1984 *The Neolithic Transition and the Genetics of Population in Europe.* Princeton University Press, Princeton.
Bar-Yosef, O.
  1981 Neolithic Sites in Sinai. In *Contributions to the Environmental History of Southwest Asia,* edited by W. Frey and H.P. Uerpmann, pp. 217-235. Wiesbaden.
  1984 Seasonality Among Neolithic Hunter-Gatherers in Southern Sinai. In *Early Herders and Their Flocks,* edited by Clutton-Brock, J. and C. Grigson, pp. 145-160. Animals and Archaeology, vol. 3. BAR International Series 202. BAR, Oxford.
  1985 The Stone Age of the Sinai Peninsula. In *Studi di Palentologia in Onore di Salvatore M. Pugl si,* edited by M. Liverani, A. Palmieri and P. Peroni, pp. 107-122. Universita di Roma "La Sapienza", Roma.
  1989 The PPNA in the Levant - an Overview. *Paléorient* 15(1):57-63.
Bar-Yosef, O. and A. Belfer-Cohen
  1989a The Origins of Sedentism and Farming Communities. *Journal of World Prehistory* 3(4):447-498.
  1989b The Levantine "PPNB" Interaction Sphere. In *People and Culture in Change: Proceedings of the Second Symposium on Upper Paleolithic, Mesolithic, and Neolithic Populations of Europe and the Mediterranean Basin,* edited by I. Hershkovitz, pp. 59-72. BAR International Series 508. BAR, Oxford.
  1991 From Sedentary Hunter-Gatherers to Territorial Farmers in the Levant. In *Between Bands and States,* edited by S.A. Gregg. Center for Archaeological Investigations Occasional Paper 9. Southern Illinois University Press, Carbondale.
Bar-Yosef, O. and M.E. Kislev
  1989 Early Farming Communities in the Jordan Valley. In *Foraging and Farming: the Evolution of Plant Exploitation,* edited by D.R. Harris and G.C. Hillman, pp. 632-642. Unwin and Hyman, London.
  1989 The Pre-Pottery Neolithic B Period in Eastern Jordan. *Paléorient* 15(1):150-156.
Binford, L.R.
  1982 The Archaeology of Place. *Journal of Anthropological Archaeology* 1:5-31.
Bogucki, P.
  1987 The Establishment of Agrarian Communities on the North European Plain. *Current Anthropology* 28:1-24.
  1988 *Forest Farmers and Stockherders: Early Agriculture and its Consequences in North-Central Europe.* Cambridge University Press, New York.
Byrd, B.F.
  in press *The Neolithic Village of Beidha: Organization and Occupation History.* Jutland Archaeological Society, vol. 23. Aarhus, Denmark.
Byrd, B.F. and E.B. Banning
  1988 Southern Levantine Pier Houses: Intersite Architectural Patterning During the Pre-Pottery Neolithic B. *Paléorient* 14(1):65-72.
Cauvin, J.
  1978 *Les Prémiers Villages de Syrie-Palestine du IX-au VII Millénaire Avant J-C.* Série Archaéologique 3, Collection de la Maison de L'Orient Ancien 4. Maison de L'Orient, Lyon.
  1987 Chronologies Relative et Absolue Dans le Néolithique du Levant Nord et D'Anatolie Entre 10,000 et 8,000 B.P. In *Chronologies du Proche Orient,* edited by O. Aurenche, J. Evin, and F. Hours, pp. 325-342. BAR International Series 379, BAR, Oxford.

1989 La Néolithisation du Levant, Huit ans Aprés. *Paléorient* 15(1):177-178.

Cauvin, J., M.C. Cauvin and O. Stordeur

    1979 Recherches Préhistorique a El Kowm (Syrié), Prémiere Campagne 1978. *Cahiers de L'Euphrate* 2, pp. 80-117. CNRS, Paris.

de Contenson, H., M.-C. Cauvin, W. Van Zeist, J.A.H. Bakker-Heeres, and A. Leroi-Gourhan

    1979 Tell Aswad (Damascene). *Paléorient* 5:153-176.   ·

Davis, S.

    1982 Climatic Change and the Advent of Domestication: the Succession of Ruminant Artiodactyls in the Late Pleistocene-Holocene in the Israel Region. *Paléorient* 8(2): 5-15.

Dayan, T., E. Tchernov, O. Bar-Yosef, and Y. Tom-Tov

    1986 Animal Exploitation in Ujrat el Mehed, a Neolithic Site in Southern Sinai. *Paléorient* 12(2):105-116.

Donaldson, M.L.

    1986 Subsistence Variation and Agriculture in the Early Levantine Neolithic. Paper Presented at the Society of American Archaeology Annual Meeting, New Orleans.

Dornemann, R.H.

    1986 *A Neolithic Village at Tell el Kowm in the Syrian Desert.* Studies in Ancient Oriental Civilization 43. University of Chicago Oriental Institute, Chicago.

Flannery, K.V.

    1969 Origins and Ecological Effects of Early Domestication in Iran and the Near East. In *The Domestication and Exploitation of Plants and Animals*, edited by P.J. Ucko and G.W. Dimbleby, pp. 73-100. Aldine, London.

Garfinkel, Y., I. Carmi, and J.C. Vogel

    1987 Dating of Horsebean and Lentil Seeds From the Pre-Pottery Neolithic B Village of Yiftah'el. *Israel Exploration Journal* 37:40-42.

Garrard, A.N., B. Byrd, P. Harvey, and F. Hivernel

    1985 Prehistoric Environment and Settlement in the Azraq Basin. A Report on the 1982 Survey Season. *Levant* 17:1-28.

Garrard, A.N., B. Byrd, and A. Betts

    1986 Prehistoric Environment and Settlement in the Azraq Basin: an Interim Report on the 1984 Excavation Season. *Levant* 18:5-24.

Garrard, A.N., A. Betts, B. Byrd, and C. Hunt

    1987 Prehistoric Environment and Settlement in the Azraq Basin: an Interim Report in the 1985 Excavation Season. *Levant* 19:5-25.

Garrard, A.N., A. Betts, B. Byrd, S. Colledge, and C. Hunt

    1988a Summary of Paleoenvironmental and Prehistoric Investigations in the Azraq Basin. In *The Prehistory of Jordan*, edited by A.N. Garrard and H.G. Gebel, pp. 311-337. BAR International Series 396. BAR, Oxford.

Garrard, A.N., S. Colledge, C. Hunt, and R. Montague

    1988b Environment and Subsistence During the Late Pleistocene and Early Holocene in the Azraq Basin. *Paléorient* 14(2):40- 49.

Goring-Morris, A.N., and A. Gopher

    1983 Nahal Issaron: a Neolithic Settlement in the Southern Negev. *Israel Exploration Journal* 33(3):149-162.

Hastorf, C.A., and V.S. Popper (editors)

    1988 *Current Paleoethnobotany.* The University of Chicago Press, Chicago.

Helmer, D.

    1989 Le Dévelopement de la Domestication au Proche-Orient de 9500 à 7500 B.P.: les Nouvelles Données D'El Kowm et de Ras Shamra. *Paléorient* 15(1):111-121.

Helms, S.W., and A.V.G. Betts

    1987 The Desert "Kites" of the Badiyat esh-Sham and North Arabia. *Paléorient* 13(1):41-67.

Henry, D.O.

    1982 The Prehistory of Southern Jordan and Relationships With the Levant. *Journal of Field Archaeology* 9:417-444.

Hillman, G.C., and M.S. Davies
  1990 Measured Domestication Rates in Wild Wheats and Barley Under Primitive Cultivation and Their Archaeological Implications. *Journal of World Prehistory* 4(2):157-222.
Hillman, G.C., S.M. Colledge, and D.R. Harris
  1989 Plant-Food Economy During the Epipaleolithic Period at Tell Abu Hureyra, Syria: Dietary Diversity, Seasonality and Modes of Exploitation. In *Foraging and Farming: the Evolution of Plant Exploitation*, edited by D.R. Harris, and G.C. Hillman, pp. 240-268. Unwin and Hyman, London.
Hole, F.
  1984 A Reassessment of the Neolithic Revolution. *Paléorient* 10(2):49-60.
  1987 Settlement and Society in the Village Period. In *The Archaeology of Western Iran*, edited by F. Hole, pp. 79-106. Smithsonian Institution, Washington D.C.
Hopf, M.
  1983 Jericho Plant Remains. In *Excavations at Jericho V*, edited by K.M Kenyon and T.A. Holland, pp. 576-621. British School of Archaeology in Jerusalem, London.
Horwitz, L.K.
  1987 The Fauna From the PPNB Site of Yiftahel: New Perspectives on Domestication. *Mitekufat Haeven* 20:181-182.
Kirkbride, D.
  1966 Beidha: An Early Neolithic Village in Jordan. *Archaeology* 19(3):199-207.
  1978 The Neolithic in Wadi Rumm: 'Ain Abu Nekheileh. In *Archaeology in the Levant*, edited by R.M. Moorey and P. Parr, pp. 1-10. Aris and Phillips, Warminster, England.
Kislev, M.E.
  1984 Emergence of Wheat Agriculture. *Paléorient* 10(2):61-70.
  1985 Early Neolithic Horsebean From Yiftah'el, Israel. *Science* 228:319-320.
  1989 Pre-Domesticated Cereals in the Pre-Pottery Neolithic A Period. In *Man and Culture in Change: Proceedings of the Second Symposium on Upper Paleolithic, Mesolithic, and Neolithic Populations of Europe and the Mediterranean Basin*, edited by I. Hershkovitz, pp. 147-151. BAR International Series 508, Oxford.
Kislev, M.E. and O. Bar-Yosef.
  1988 The Legumes: The Earliest Domesticated Plants in the Near East. *Current Anthropology* 29(1):175-179.
Kislev, M.E., O. Bar-Yosef, and A. Gopher.
  1986 Early Neolithic Domesticated and Wild Barley From the Netiv Hagdud Region in the Jordan Valley. *Israel Journal of Botany* 35:197-201.
Kohler Rollefsion, I.
  1989 Changes in Goat Exploitation at 'Ain Ghazal Between the Early and Late Neolithic: a Metrical Analysis. *Paléorient* 15(1):141-146.
Ladizinsky, G.
  1989 Origin and Domestication of the Southwest Asian Grain Legumes. In *Foraging and Farming: the Evolution of Plant Exploitation*, edited by D.R. Harris and G.C. Hillman, pp. 374-389. Unwin and Hyman, London.
Ladizinsky, G., and A. Adler.
  1976 The Origin of Chickpea, Cicer Arietinum L. *Euphytica* 25:211-217.
Legge, A.J., and P.A. Rowley-Conwy
  1987 Gazelle Killing in Stone Age Syria. *Scientific American* 257(8):88-95.
Lewthwaite, J.
  1986 The Transition to Food Production: a Mediterranean Perspective. In *Hunters in Transition: Mesolithic Societies of Temperate Eurasia and Their Transition to Farming*, edited by M. Zvelebil, pp. 53-66. Cambridge University Press, Cambridge.
Marks, A.E.
  1977 Prehistoric Settlement Patterns in the Avdat/Aqev Area. In *Prehistory and Paleoenvironments in the Central Negev, Israel, Volume 2*, edited by Marks, A.E., pp. 131-158. Southern Methodist University, Dallas.

McCorriston, J., and F. Hole.
　1991　The Ecology of Seasonal Stress and the Origins of Agriculture in the Near East. *American Anthropologist* 93 (1):46-69.

Miller, N.F.
　1991　The Near East. In *Progress in Old World Paleoethnobotany,* edited by W. van Zeist, K. Wasylikowa, and K.E. Behre, pp. 133-160. A.A. Balkema, Rotterdam.

Moore, A.M.T.
　1985　The Development of Neolithic Societies in the Near East. *Advances in World Archaeology* 4:1-69.
　1989　The Transition From Foraging to Farming in Southwest Asia: Present Problems and Future Directions. In *Foraging and Farming: the Evolution of Plant Exploitation,* edited by D.R. Harris and G.C. Hillman, pp. 620-631. Unwin Hyman, London.

Nissen, H.J., M. Muheisen, and H. G. Gebel.
　1987　Report on the First Two Seasons of Excavations at Basta (1986-1987). *Annual of the Department of Antiquities of Jordan* 31:79-119.

Noy, T.
　1989　Gigal I. A Pre-Pottery Neolithic Site, Israel. The 1985- 1987 Seasons. *Paléorient* 15(1):15-22.

Pinkas, R., D. Zamir and G. Ladizinsky.
　1985　Allozyme Divergence and Evolution in the Genus *Lens. Plant Systematics and Evolution* 151(131):140-140.

Rollefson, G.O.
　1989　The Late Aceramic Neolithic of the Levant: a Synthesis. *Paléorient* 15(1):168-173.

Servello, A.F.
　1976　Nahal Divshon: a Pre-Pottery Neolithic B Hunting Camp. In *Prehistory and Paleoenvironments in the Central Negev, Israel,* vol. 1, edited by A.E. Marks, pp. 349-371. Southern Methodist University, Dallas.

Sheratt, A.G.
　1980　Water, Soil, and Seasonality in Early Cereal Cultivation. *World Archaeology* 11:313-330.

Simmons, A.H.
　1981　A Paleosubsistence Model for Early Neolithic Occupation of the Western Negev Desert. *Bulletin of the American Oriental Society* 242(31):31-50.

Simmons, A.H., I. Kohler-Rollefson, G.O. Rollefson, R. Mandel, and Z. Kafafi.
　1988　`Ain Ghazal: a Major Neolithic Settlement in Central Jordan. *Science* 240:35-39.

Smith, P., O. Bar-Yosef, A. Sillen.
　1984　Archaeological and Skeletal Evidence for Dietary Change During the Late Pleistocene-Early Holocene in the Levant. In *Paleopathology at the Origins of Agriculture,* edited by M. Cohen and G. Armelagos, pp. 101-137. Academic Press, New York.

Speth, J.D.
　1983　*Bison Kills and Bone Counts: Decision Making by Ancient Hunters.* University of Chicago Press, Chicago.

Stordeur, D.
　1989　El Kowm 2 Caracol et le PPNB. *Paléorient* 15(1):102-110.

Tchernov, E. and O. Bar-Yosef.
　1982　Animal Exploitation in the Pre-Pottery Neolithic B Period at Wadi Tbeik, Southern Sinai. *Paléorient* 8(2):17-37.

van Zeist, W.
　1972　Palaeobotanical Results of the 1970 Season at Cayonu, Turkey. *Helinium* 12:3-19.
　1986　Plant Remains from Neolithic el Kowm, Central Syria. In *A Neolithic Village at Tell el Kowm in the Syrian Desert,* edited by R.H. Dornemann, pp. 65-68. University of Chicago Oriental Institute, Chicago.
　1988　Some Aspects of Early Neolithic Plant Husbandry in the Near East. *Anatolica* 15:49-68.

van Zeist, W., and J.A.H. Bakker-Heeres.
　1982　Archaeobotanical Studies in the Levant: Neolithic Sites in the Damascus Basin: Aswad, Ghoraife, Ramad. *Palaeohistoria* 24:165-256.
　1984a　Archaeobotanical Studies in the Levant 3: Late Paleolithic Mureybet. *Palaeohistoria* 26:171-199.

1984b Archaeobotanical Studies in the Levant 2. Neolithic and Halaf Levels at Ras Shamra. *Palaeohistoria* 26:151-170.

van Zeist, W., and W. Waterbolk-van Rooijen.

1985 The Palaeobotany of Tell Bouqras, Eastern Syria. *Paléorient* 11(2):131-147.

Voigt, M.M.

1983 *Hajji Firuz Tepe, Iran: the Neolithic Settlement.* Hasanlu Excavation Reports, 1. University Museum Monograph 50. University of Pennsylvania, Philadelphia.

Zarins, J.

1989 Jebel Bishri and the Amorite Homeland: the PPNB Phase. In *To the Euphrates and Beyond*, edited by O. Chaex, H. Curvers, and P. Akkermans, pp. 29-51. A.A. Balkema, Rotterdam.

1990 Early Pastoral Nomadism and the Settlement of Lower Mesopotamia. *Bulletin of the American Schools of Oriental Research* 280:31-65.

Zohary, D.

1989 Domestication of the Southwest Asian Neolithic Crop Assemblage of Cereals, Pulses, and Flax: The Evidence From the Living Plants. In *Foraging and Farming: The Evolution of Plant Exploitation*, edited by D.R. Harris and G.C. Hillman, pp. 358-373. Unwin Hyman, London.

Zohary, D., and I. Hopf

1988 *Domestication of Plants in the Old World.* Clarendon Press, Oxford.

Zvelebil, M.

1989 On the Transition to Farming in Europe, or What was Spreading With the Neolithic: a Reply to Ammerman. *Antiquity* 63:379-383.

Zvelebil, M., and P. Rowley-Conwy

1986 Foragers and Farmers in Atlantic Europe. In *Hunters in Transition: Mesolithic Societies of Temperate Eurasia and Their Transition to Farming*, edited by M. Zvelebil, pp. 67- 94. Cambridge University, Cambridge.

5

The Beginnings of Food Production in the Eastern Sahara

Angela E. Close
Fred Wendorf
Southern Methodist University

In northeastern Africa, the last period of increased rainfall in the Upper Pleistocene was associated with the Middle Palaeolithic and dates to at least 50,000 years ago and probably considerably more (Wendorf et al. 1991). After that, there was nothing and no one in the desert, except wind, sand and stars (Figure 1), until the Early Holocene, when the region was colonized by cattle pastoralists. This chapter deals with the beginnings of food production in the Holocene Eastern Sahara, but the antecedents of this process lie in the Nile Valley during the last stages of the Pleistocene.

Only the Nile Valley was inhabitable during the late Pleistocene and even here, after the maximum of the Last Glaciation, the Nile was a very different river from that of today, with less water flowing through a multitude of braided channels across a high, wide floodplain. The Valley was occupied by Late Palaeolithic groups, who seem to have subsisted primarily upon floodplain plant foods, supplemented by a markedly seasonal exploitation of catfish. Mammals—wild cattle, hartebeest, gazelle, and occasional wild ass and hippo—were also taken, although they provided relatively minor parts of the diet (Wendorf et al. 1988, 1989).

At the end of the Pleistocene, there were profound —perhaps even catastrophic—changes in the regimen of the Nile. At about 13,000 B.P., increased rainfall in the headwaters region resulted in extremely high floods, after which the river began to downcut in a single channel within a much reduced floodplain. Unfortunately, very little is known about the archaeology of the period immediately following this change. However, the narrower channel would have greatly reduced the availability of the most important Late Palaeolithic plant foods and probably of catfish, which were formerly caught while spawning in the wide embayments of the Late Palaeolithic floodplain.

Holocene Colonization of the Eastern Sahara

In the Eastern Sahara, the rains came again around the beginning of the Holocene, and at least most of the rainfall resulted from a northward shift of the southern (monsoon) rainfall belts. The actual beginning of the Holocene wet period is not well dated. Interpreted conservatively, the available data indicate that the rains began in the southern part of the Eastern Sahara before 11,000 B.P. However, there is essentially no trace of human occupation much before 9500 B.P. Thus, between the onset of the rains and the arrival of the people, there seems to have been a lag of some 1500 years and possibly more.

By 9500 B.P., human groups had begun to move into the southern Eastern Sahara, where they are best known from sites associated with playas in the regions of Gebel Nabta (100 km west of Abu Simbel) and Bir Kiseiba (200 km west of Abu Simbel) (Figure 2) (Wendorf and Schild 1980; Wendorf et al. 1984). In the southern part of the desert, at least, they probably

Figure 1. The southern Eastern Sahara—rainless today, as it was before the beginning of the Early Holocene wet period.

came from the east—from the Nile Valley—primarily because they seem not to have come from anywhere else. There is no evidence for earlier occupation in the desert to the west, or in the desert itself or along the coast to the north. Since the rains moved into the desert from the south, it would seem reasonable that people should have advanced with them, but there is no positive evidence that they did. Extensive research in northern Sudan has not revealed any sites older than 9000 B.P. and very few earlier than 8000 B.P. (Kuper 1981, 1986, 1988; Richter 1989; Schuck 1988, 1989). By process of elimination, we therefore arrive at the conclusion that the first human colonists of the Holocene Eastern Sahara came in from the east.

Argument for Domestic Status of Cattle

We believe that even the first Holocene colonists of the desert brought domestic cattle with them, although this has not been universally accepted (Clutton-Brock 1989; Muzzolini 1989b; Smith 1984). Every site dating earlier than 8000 B.P. (this is also true of sites post-dating 8000 B.P., but that is less controversial) from which there is a moderately large faunal collection has

yielded a few bones of large bovids. The largest collection without large bovids has 41 identifiable mammal bones (Wendorf et al. 1987: Table 1). Large bovids found in Holocene contexts in northeastern Africa may be African buffalo (*Syncerus caffer*), giant buffalo (*Pelorovis antiquus*), wild cattle (*Bos primigenius*), or domestic cattle (*B. primigenius* f. taurus). The specimens from the Sahara fall within the size-range of African buffalo, but differ morphologically from that form (Gautier 1984: 61; this particular comparison has been re-examined in detail by J. Peters and still found inapplicable), and they differ both morphologically and osteometrically from the giant buffalo (Gautier 1984). They therefore pertain to wild or domestic cattle. In terms of size, they could be either wild or domestic; however, size is a treacherous criterion and our argument that they were domestic is based primarily on ecological considerations.

The evidence for environmental conditions in the Eastern Sahara before about 8000 B.P. is incomplete, but relatively consistent. Most of the published data from floral remains are based upon Neumann's (1989) identifications of charcoal. The samples which have been directly dated to before 8000 B.P. indicate the presence of tamarisk and acacia in the Sand Sea and at

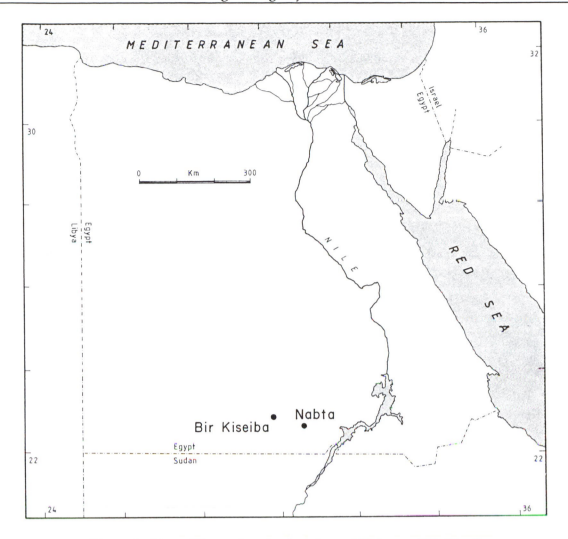

Figure 2. Map of Egypt showing locations of Nabta and Bir Kiseiba.

Abu Ballas. The latter area also has a few traces of chenopods, *Maerua*, and *Leptadenia* (Neumann 1989: Tables 1-2a); the last two are southern forms.

Recent work at Site E-75-6 at Nabta has yielded a wide variety of carbonized plant remains which are not yet finally identified. Thus far, only the material recovered in 1990 has received more than preliminary sorting. The types identified to date are listed in Table 1 (Wasylikowa 1991). The collection is dominated by *Zizyphus* stones, grass grains, leguminous plant seeds and probably seeds of the mustard and caper families; the unidentified material consists of many examples of a few types.

Ecological interpretation of this material is not yet firm, since several genera have not yet been identified and there are no identifications to the species level, but some observations can be made. There are no fruits or seeds of aquatic plants *sensu stricto*, confirming the ephemeral nature of the playa lake. The *Scirpus/*

Schoenoplectus and *Cyperus/Fuirena* types could have grown in a reed swamp or sedge-belt at the playa's edge; *Panicum* (if it is *P. repens*) and *Echinochloa* also would grow on the damp shore, and the tubers are likely to be derived from a similar environment.

Only one tree, *Zizyhpus* sp., is definitely present in the macrofossils, although *Capparis* is possible; *Tamarix* and several species of *Acacia* have been identified among the wood charcoal (Hala Barakat in litt. 1991). *Zizyphus* is a Sahelo-Saharan element, but also grows in the desert zone (Ritchie 1987). The abundance of its fruits suggests that the trees were easily accessible to the inhabitants of E-75-6.

The abundance and diversity of plant remains indicate that the local vegetation was rich in species and, in some seasons, relatively luxuriant. Sedges and some grasses favoring wet habitats grew around the playa, a contracted vegetation with *Zizyphus* trees and perennial grasses grew where surface or ground-water was

Table 1. Preliminary identifications of plant taxa at Site E-75-6 (Nabta)

	No. of samples where taxon occurs
Trees	
Zizyphus sp.	51
Grasses (sub-family Panicoideae, tribe Paniceae)	
Echinochloa type	18
Digitaria type	3
Urochloa/Brachiaria type	5
Panicum type	21
Setaria type	12
(sub-family Panicoideae, tribe Andropogoneae)	
Sorghum sp.	30
Other herbs or shrublets	
Cyperus/Fuirena type	5
Scirpus/Schoenoplectus type	5
Rumex type	1
Cucurbitaceae	23
Arnebia type	25
Unidentified	
Leguminosae, several types	frequent
Cruciferae?, one type	frequent
Chenopodiaceae?	2
Malvaceae?	1
Capparidaceae?	frequent
Tubers (unidentified parenchymatous tissues)	
Several unidentified fruits and seeds	

Figure 3. Landscape near El Fasher, in northwestern Sudan, which is probably comparable to the Eastern Sahara during the Early Holocene.

available for most of the year, while annual grasses developed after the summer rains as a relatively luxuriant *acheb* vegetation (Wasylikowa 1991) (Figure 3).

For the plains, which include the Nabta and Kiseiba regions, Neumann has suggested that the rainfall during this early period was only some 20–50 mm per annum (1989: 142), which seems somewhat low. She further suggests that temperatures were cooler (which would reduce the loss of water by evaporation) because of the virtual absence of southern species, and that temperatures did not increase until about 7000 B.P. However, even according to her own data, southerly plants were appearing at Abu Ballas in some frequency by 7500 B.P. (1989: Table 2b) and had been present, even if rare, by 8300 B.P. At Nabta, southern species are present before 8000 B.P. (Table 1). It is therefore possible that warming may have begun well before 8000 B.P. Since this would have resulted in a higher rate of evaporation, the annual rainfall may have been somewhat higher than Neumann's figure.

A figure of ca. 200 mm of rainfall per annum has previously been estimated on the basis of the fauna (Wendorf et al. 1984: 408). The indisputably wild, herbivore faunas (excluding micro-mammals) seem to consist almost entirely of hare and gazelle (most of them small), indicating a dry environment, since these desert-adapted forms need to drink only rarely, if at all (Gautier 1980, 1984; Van Neer and Uerpmann 1989). Other, extremely rare forms include oryx from the Sand Sea (Van Neer and Uerpmann 1989: 326). Since oryx runs with gazelle (Gautier 1984: 59) and also does not need to drink (Van Neer and Uerpmann 1989: 326), this also suggests a dry environment. One fragment of giraffe bone is known from Abu Ballas (Van Neer and Uerpmann 1989: 321, 328), but these animals can wander over long distances after the rains and drink only infrequently; evidence from the Gilf (although mostly from later periods) suggests that there might have been a resident giraffe population there (Peters 1988: 76). The Gilf el Kebir-Uweinat area, like the massifs of the Central Sahara, received more (orographic) rainfall than the plains to the east and probably supported a somewhat wider range of animals (Peters 1987). The fragmentary elephant tooth from Bir Kiseiba was probably a fossil (Gautier 1984: 56-57).

The wild herbivores were therefore essentially limited to hare and small gazelle, and found associated with cattle. Large wild bovids are known from earlier periods in the Sahara, but with a full size-range of associated species. For example, the giant buffalo with the Last Interglacial Middle Palaeolithic at Bir Tarfawi was found with a wide variety of gazelles and antelopes, giraffe, camel, ass(?), and rhinoceros (Gautier in press). A fauna consisting of hare, small gazelle, and cattle, with no intermediate forms, would be highly improbable ecologically. This is made particularly clear by the absence of hartebeest, which is always present with (wild) cattle in the Nile Valley and which is actually better adapted to arid conditions than are cattle.

There was also the problem of surface water. Cattle are thirsty creatures and must drink every day if they are to thrive, or every other day if they are simply to survive. They would not occur naturally in an environment which, even generously estimated, had no more than 200 mm of rainfall and which supported only hare, gazelle, and the occasionally oryx—all of them essentially non-drinkers. There was no permanent, standing water in the southern part of the Eastern Sahara during the Holocene (there had been permanent lakes during the much wetter Last Interglacial), only ephemeral playa lakes. These would have filled during the summer rains, but then have stood dry for much of the year. During most seasons, the only way to obtain water would have been to dig wells. We may therefore reject the idea that the cattle found in this area in the Early Holocene could have been wild.

Domestication as an Aspect of Human Expansion

If the colonists of the Eastern Sahara came from the Nile Valley, bringing domestic cattle with them, then we might conclude that the actual process of domestication took place at the end of the Pleistocene or beginning of the Holocene in the Nile Valley. The wild progenitors of domestic cattle were present in the Valley (as they were not in the desert), although all indications are that they were simply hunted for meat. There is evidence that wild cattle had a particular importance for the human inhabitants of the Valley; horn-cores were used as grave-markers at about 14,000 B.P. (Wendorf 1968a: 875), a habit which does not obviously denote a predisposition to food production. Unfortunately, the space-time continuum of this area is almost unknown archaeologically during this period and anything more is speculation.

Alternatively and more probably, domestication may have occurred as part of the process of human expansion into the Sahara in the tenth millennium B.P. — a process which was at least made easier and perhaps even made possible by the presence of domestic animals. The return of rainfall to the Early Sahara in the early Holocene opened a new region into which the occupants of the Nile Valley could expand. It was not, however, simply an extension of the Valley, but a more precarious environment, in which resources were much more thinly scattered and much less reliable. We would not, therefore, expect human adaptations to be the same in the desert as they had been in the Valley. How the desert adaptation developed is unknown, but when we first see it the relationship between people and their cattle is already one of local symbiosis (cf. Rindos 1984). Together, the two species survived in an area where one of them, and probably both of them, would have been unable to survive alone. For the cattle, their herders could force them to move often enough, far enough and, most probably, in the right direction to find sufficient grazing and open water. Even the more desert-adapted hartebeest could not manage this in the Eastern Sahara. For the people, cattle seem to have served as a reliable, on-the-hoof source of protein in a very marginal environment.

As noted above, the bones of domestic cattle are present whenever the overall faunal sample is of some size; however, they are never common. This rarity may mean the cattle were not commonly killed for meat, but were kept as living sources of blood and milk, as they are today among Subsaharan and Sahelian pastoralists. For example, milk is the basic food for the Kel Tamasheq of southern Mali, who frequently eat it with plants similar to those listed in Table 1 and, during the wet season, live on little else but milk

(Smith 1980). In contrast, in the Nile Valley at this time there is no evidence that people were doing anything with wild cattle other than hunting and killing them for meat.

We thus see three parallel dichotomies at this time in northeast Africa: the Nile vs. the Desert, wild vs. domestic cattle, and meat vs. milk and blood. This pattern agrees with the idea that domestication occurred as part of the expansion into the desert, not as part of a purely Nilotic adaptation.

Even with less than 200 mm of rain per annum, particularly if temperatures were somewhat cooler, there would have been some standing water in the Eastern Sahara during and after the rains, so that the cattle, once brought into the area, would have been able to drink during that season. During the very earliest period there is no evidence for wells; it is possible that the first occupations were only seasonal and that people returned to the Nile Valley as the surface water dried up. During the second half of the ninth millennium, large, deep wells were dug (Wendorf and Schild 1980: Figure 3.73), some of them with shallow basins at the side (Figure 4; Close 1984a: Figure 12.6). The latter is a form still used in the desert (for example, at Laqiya Arba'in on the main camel-road); water is bailed out of

Figure 4. A partially excavated Neolithic well near Bir Kiseiba; in the foreground is a shallow sub-basin like those used today for watering stock.

the well into the basin, from which the livestock drink. The wells occur in the lower parts of playa basins, indicating their use during the drier part of the year. It is possible that people at that time and their cattle were spending longer periods in the desert, and may even have been able to stay year-round, although it would be difficult to demonstrate this.

Early Holocene Diet in the Eastern Sahara

For the diet of these early colonists of the Eastern Sahara, we suppose that they drank the milk and blood of their domestic cattle. We know that they obtained additional protein from local wild animals (predominantly hare and gazelle) which were presumably hunted. On the occasional death of a domestic animal, whether killed by its herders or dying of other causes, the meat would almost certainly have been eaten. Stone grinding equipment occurs in even the earliest of these sites (Connor 1984: 239; Close 1984b: 346) and is presumed to have been used for processing plant foods, although the only direct evidence from the earliest period is a few seeds of Leguminosae from Site E-77-7, near Nabta.

The carbonized plant remains from Site E-75-6 at Nabta, which has a stratified sequence dating from about 8600 B.P. to about 7500 B.P., were found within and around burned areas in houses or shelters, where they were presumably being cooked to be eaten. They include (wild) grasses, legumes, and other plants (Table 1), which still grow in the area after rain, as well as some that today grow farther south, and sorghum and millets, which also appear to be wild (Wasylikowa 1991).

Harlan (1989a, 1989b) has summarized the ethnographic evidence for the use of wild grass seeds by Saharan and Subsaharan groups in the nineteenth and twentieth centuries. Harvested wild plants were more important and more reliable sources of food than were domesticated forms in the marginal areas—which would certainly have included the Early Holocene Eastern Sahara. Harlan also observes that wild grass seeds seem to be more nutritious than domesticated cereals (1989b: 73). The evidence from Nabta suggests that this recent and sub-recent Saharan practice may have begun by 8600 years ago.

Direct evidence for plant foods is very rare from other sites in the Nabta area and is completely lacking from sites elsewhere in the Desert. There is no evidence for the use of cereals domesticated in southwestern Asia before their first appearance in the Fayum and at Merimde shortly after 6000 B.P. (Caton Thompson and Gardner 1934; Hassan 1988). Barley and wheat have previously been reported from Early

and Middle Neolithic sites at Nabta (el Hadidi 1980: 346-348), but extensive re-excavation of both sites has failed to reconfirm the presence of cereals at the site. We therefore believe that they were not in true archaeological association. The relative aridity of the true desert where, by that time, the Holocene wet phases were drawing to a close would have hindered the spread of cereals westward from the Fayum. It is quite possible that all of the tens of thousands of grinding stones known from the Eastern Sahara were related to the use of local, or Subsaharan, plant foods.

Other Domesticates

The southwest Asian cereals became the staples of Pharaonic Egypt, presumably because of their high productivity. Thus far, the sorghums and millets used at Nabta over 8000 years ago seem to be morphologically wild and, presumably, were simply parts of an economy based upon gathering wild plant foods. However, it has been suggested that sorghum and millet were first cultivated in Ethiopia or in the savanna region (Doggett 1976; Harlan 1975); the northward extension of the monsoon zone in the early Holocene also makes southwestern Egypt a possibile area for these early cultigens. Sorghum did eventually find its way to the Nile Valley. In Pharaonic times, it may have been cultivated primarily in the south, although there is some doubt that it was cultivated at all (Darby et al. 1977: 494–496); however, it was known throughout the Valley in recent times (Täckholm et al. 1941: 520-541).

Sheep or goat occur in Middle Neolithic contexts at Nabta (E-75-8) shortly after 7000 B.P. (Gautier 1980: 332-338; their presence was confirmed by re-excavation of the site in 1990). Caprovids were domesticated in southwestern Asia (*sensu lato*) (Uerpmann 1989: 933; Hole 1989) in the ninth millennium B.P. Their wild ancestors are not known from North Africa (Richard Meadow (personal communication, 1990; Muzzolini [1989a:12–13] strongly disagrees), so that any caprovids found here must be domesticates brought in by their herders. The Middle Neolithic domestic caprovids from Nabta are among the earliest in Africa. That they spread so far so fast indicates how well they fit into an economy that was, by that time, apparently permanently desert-based.

Conclusions

Early Neolithic sites in the southern part of the Eastern Sahara, even those with radiocarbon dates earlier than 9000 B.P., usually yield bones believed to be those of domestic cattle. The age and location of the sites mean that cattle must have been independently domesticated in Africa, probably in the tenth millennium B.P.

At the end of the Pleistocene, the Nile Valley in Egypt and Sudan was a sharply defined and delimited area of inhabitable territory—it can be seen, not too inaccurately, as an extremely elongated oasis. The Valley underwent profound environmental change at that time, apparently for the worse. The inhabitants of the Valley had to adapt to these changes, since it was not possible for them to move elsewhere. However, the only human response we can see to the stress is that occasionally groups slaughtered each other (Anderson 1968; Wendorf 1968b). There is no evidence that they attempted to enhance their food resources, by food production or any other means. Instead, food production in northeastern Africa arose in connection with the opening of a new, previously uninhabited area and with the expansion of human groups into it. The process of human expansion and the process of cattle domestication were causally related in chicken-and-egg fashion. Each permitted or led to the other, and it would be hazardous and probably misleading to regard either alone as the prime mover.

Finally, the domestic cattle in the Eastern Sahara were apparently not being kept for meat, and this may imply a special form of domestication. If animals are used for meat, whether they are domestic or wild, when the time comes for them to serve their ultimate purpose, they do not cooperate, nor is there any need for their cooperation. However, when cattle are kept as providers of milk and/or blood, these resources can be collected only with the animals' cooperation. This implies a very close relationship between the people and their animals. It is not clear how much time is required for changes in bone morphology to appear as cattle become domestic. The bones recovered from the Eastern Sahara are, in any case, too few and too fragmentary to indicate their morphological status decisively. However, the animals must have been very tame and physically accustomed to their herders. Their domestication may have been qualitatively different from the carnivorous variety more common in southwestern Asia.

References Cited

Anderson, J. E.
   1968 Late Paleolithic Skeletal Remains from Nubia. In *The Prehistory of Nubia*, edited by F. Wendorf, pp. 996-1040. Fort Burgwin Research Center and Southern Methodist University Press, Dallas.
Caton Thompson, G., and E. W. Gardner
   1934 *The Desert Fayum.* Royal Anthropological Institute, London.
Close, A. E.
   1984a Report on Site E-80-1. In *Cattle-Keepers of the Eastern Sahara: The Neolithic of Bir Kiseiba*, assembled by F. Wendorf and R. Schild, edited by A. E. Close, pp. 251-297. Department of Anthropology, Southern Methodist University, Dallas.
   1984b Report on Site E-80-4. In *Cattle-Keepers of the Eastern Sahara: The Neolithic of Bir Kiseiba*, assembled by F. Wendorf and R. Schild, edited by A. E. Close, pp. 325-349. Department of Anthropology, Southern Methodist University, Dallas.
Clutton-Brock, J.
   1989 Cattle in Ancient North Africa. In *The Walking Larder: Patterns of Domestication, Pastoralism and Predation*, edited by J. Clutton-Brock, pp. 200-206. Unwin Hyman, London.
Connor, D. R.
   1984 Report on Site E-79-8. In *Cattle-Keepers of the Eastern Sahara: The Neolithic of Bir Kiseiba*, assembled by F. Wendorf and R. Schild, edited by A. E. Close, pp. 217-250. Department of Anthropology, Southern Methodist University, Dallas.
Darby, W. J., P. Ghalioungui, and L. Grivetti
   1977 *Food: The Gift of Osiris.* Academic Press, London.
Doggett, H.
   1976 Sorghum: *Sorghum bicolor* (*Gramineae - Andropogoneae*). In *Evolution of Crop Plants*, edited by N. W. Simmonds, pp. 112-117. Longman, London.
Gautier, A.
   1980 Contributions to the Archaeozoology of Egypt. In *Prehistory of the Eastern Sahara*, by F. Wendorf and R. Schild, pp. 317-344. Academic Press, New York.

1984 Archaeozoology of the Bir Kiseiba Region, Eastern Sahara. In *Cattle-Keepers of the Eastern Sahara: The Neolithic of Bir Kiseiba*, assembled by F. Wendorf and R. Schild, edited by A. E. Close, pp. 49-72. Department of Anthropology, Southern Methodist University, Dallas.

1992 The Middle Paleolithic Archaeofaunas from Bir Tarfawi (Western Desert, Egypt). In *The Prehistory of Bir Tarfawi and Bir Sahara East*, assembled by F. Wendorf and R. Schild, edited by A. E. Close. Southern Methodist University Press, Dallas.

el Hadidi, M. N.

1980 Vegetation of the Nubian Desert (Nabta region). In *Prehistory of the Eastern Sahara*, by F. Wendorf and R. Schild, pp. 345-351. Academic Press, New York.

Harlan, J. R.

1975 *Crops and Man*. American Society of Agronomy, Madison, Wisconsin.

1989a Wild-grass Seed Harvesting in the Sahara and Sub-Sahara of Africa. In *Foraging and Farming. The Evolution of Plant Exploitation*, edited by D. R. Harris and G. C. Hillman, pp. 79-98. Unwin Hyman, London.

1989b Wild Grass Seeds as Food Sources in the Sahara and Sub-Sahara. *Sahara* 2:69-74.

Hassan, F. A.

1988 The Predynastic of Egypt. *Journal of World Prehistory* 2:135-185.

Hole, F.

1989 A Two-part, Two-stage Model of Domestication. In *The Walking Larder: Patterns of Domestication, Pastoralism and Predation*, edited by J. Clutton-Brock, pp. 97-104. Unwin Hyman, London.

Kuper, R.

1981 Untersuchungen zur Besiedlungsgeschichte der östlichen Sahara: Vorbericht über die Expedition 1980. *Beiträge zur allgemeinen und vergleichenden Archäologie* 3:215-275.

1986 Wadi Howar and Laqiya—Recent Field Studies into the Early Settlement of Northern Sudan. In *Nubische Studien*, edited by M. Krause, pp. 129-136. Verlag Philipp von Zabern, Mainz.

1988 Neuere Forschungen zur Besiedlungsgeschichte der Ost-Sahara. *Archäologisches Korrespondenzblatt* 18:127-142.

Muzzolini, A.

1989a Les débuts de la domestication des animaux en Afrique: faits et problemes. *Ethnozootechnie* 42:7-22.

1989b La "néolithisation" du Nord de l'Afrique et ses causes. In *Néolithisations*, edited by O. Aurenche and J. Cauvin, pp. 145-186. BAR International Series 516. BAR, Oxford.

Neumann, K.

1989 Zur Vegetationsgeschichte der Östsahara im Holozän: Holzkohlen aus prähistorischen Fundstellen. In *Forschungen zur Umweltgeschichte der Östsahara*, edited by R. Kuper, pp. 13-181. Africa Praehistorica 2. Heinrich-Barth Institut, Köln.

Peters, J.

1987 The Faunal Remains Collected by the Bagnold-Mond Expedition in the Gilf Kebir and Jebel Uweinat in 1938. *Archéologie du Nil Moyen* 2:251-264.

1988. The Palaeoenvironment of the Gilf Kebir-Jebel Uweinat Area During the First Half of the Holocene: the Latest Evidence. *Sahara* 1:73-76.

Richter, J.

1989 Neolithic Sites in the Wadi Howar (Western Sudan). In *Late Prehistory of the Nile Basin and the Sahara*, edited by L. Krzyzaniak and M. Kobusiewicz, pp. 431-442. Poznan Archaeological Museum, Poznan.

Rindos, D.

1984 *The Origins of Agriculture: An Evolutionary Perspective*. Academic Press, Orlando.

Ritchie, J. C.

1987 A Holocene Pollen Record from Bir Atrun, Northwest Sudan. *Pollen et Spores* 29:391-410.

Schuck, W.

1988 Wadi Shaw—eine Siedlungskammer im Nord-Sudan. *Archäologisches Korrespondenzblatt* 18:143-153.

1989 From Lake to Well: 5000 Years of Settlement in Wadi Shaw (northern Sudan). In *Late Prehistory of the Nile Basin and the Sahara*, edited by L. Krzyzaniak and M. Kobusiewicz, pp. 421-429. Poznan Archaeological Museum, Poznan.

Smith, A. B.

 1984 The Origins of Food Production in Northeast Africa. In *Palaeoecology of Africa and the Surrounding Islands*, vol. 16, edited by J. A. Coetzee and E. M. Van Zinderen Bakker, pp. 317-324. Balkema, Rotterdam.

Smith, S. E.

 1980 The Environmental Adaptation of Nomads in the West African Sahel: a Key to Understanding Prehistoric Pastoralists. In *The Sahara and the Nile: Quaternary Environments and Prehistoric Occupation in Northern Africa*, edited by M. A. J. Williams and H. Faure, pp. 467-487. Balkema, Rotterdam.

Täckholm, V., G. Täckholm, and M. Drar

 1941 *Flora of Egypt*, vol. 1. Fouad I University, Cairo.

Uerpmann, H.-P.

 1989 Animal Exploitation and the Phasing of the Transition from the Palaeolithic to the Neolithic. In *The Walking Larder: Patterns of Domestication, Pastoralism and Predation*, edited by J. Clutton-Brock, pp. 91-96. Unwin Hyman, London.

Van Neer, W., and H.-P. Uerpmann

 1989 Palaeoecological Significance of the Holocene Faunal Remains of the B.O.S.-Missions. In *Forschungen zur Umweltgeschichte der Ostsahara*, edited by R. Kuper, pp. 307-341. Africa Praehistorica 2. Heinrich-Barth-Institut, Köln.

Wasylikowa, K.

 1991 Seeds and Fruits from Nabta Playa from the Field Seasons 1990 and 1991. Ms. on file, Southern Methodist University, Dallas.

Wendorf, F.

 1968a Late Paleolithic Sites in Egyptian Nubia. In *The Prehistory of Nubia*, edited by F. Wendorf, pp. 791-953. Fort Burgwin Research Center and Southern Methodist University Press, Dallas.

 1968b Site 117: a Nubian Final Paleolithic Graveyard near Jebel Sahaba, Sudan. In *The Prehistory of Nubia*, edited by F. Wendorf, pp. 954-995. Fort Burgwin Research Center and Southern Methodist University Press, Dallas.

Wendorf, F., A. E. Close, and R. Schild

 1987 Early Domestic Cattle in the Eastern Sahara. In *Palaeoecology of Africa and the Surrounding Islands*, Vol. 18, edited by J. A. Coetzee, pp. 441-448. Balkema, Rotterdam.

Wendorf, F., A. E. Close, R. Schild, A. Gautier, H. P. Schwarcz, G. H. Miller, K. Kowalski, H. Krolk, A. Bluszcz, D. Robins, R. Grün, and C. McKinney

 1992 Chronology and Stratigraphy of the Middle Paleolithic at Bir Tarfawi, Egypt. In *The Origins of Culture and the Earliest Industries of Africa*, edited by J. D. Clark. Römisch-Germanisches Zentralmuseum, Mainz, in press.

Wendorf, F. and R. Schild

 1980 *Prehistory of the Eastern Sahara*. Academic Press, New York.

Wendorf, F., R. Schild (assemblers), and A. E. Close (editor)

 1984 *Cattle-Keepers of the Eastern Sahara: The Neolithic of Bir Kiseiba*. Department of Anthropology, Southern Methodist University, Dallas.

 1989 *The Prehistory of Wadi Kubbaniya*. 2 vols. Southern Methodist University Press, Dallas.

Wendorf, F., R. Schild, A. E. Close, G. C Hillman, A. Gautier, W. Van Neer, D. J. Donahue, A. J. T. Jull, and T. W. Linick

 1988 New Radiocarbon Dates and Late Palaeolithic Diet at Wadi Kubbaniya, Egypt. *Antiquity* 62:279-283.

6

Desperately Seeking Ceres: A Critical Examination of Current Models for the Transition to Agriculture in Mediterranean Europe

Randolph E. Donahue
University of Sheffield, England

The purpose of this paper is to briefly review current models for the transition to agriculture in Mediterranean Europe with respect to recently collected and re-evaluated archaeological data. The results of this review suggest that a major revision in present thinking is required concerning the process and the participants in the transition. No longer is there reason to argue that neolithic colonists from the eastern Mediterranean played a significant part. Instead, there is increasing evidence for the active and effective role of mesolithic hunter-gatherers who experienced gradual subsistence changes prior to and associated with the introduction of domesticated animals and plants and new technologies, eventually leading toward fully agricultural economies.

A Review of Current Models

During most of this century the neolithic of Mediterranean Europe was viewed in diffusionist terms within a culture-historical framework. Although rarely explicit, the underlying cause for the diffusion of material culture and for the appearance of advanced agricultural economies was migration; colonists arrived in Italy and the western Mediterranean from the eastern Mediterranean, Anatolia, or the Near East (Bernabó Brea 1956; Laviosa Zambotti 1943). In the past 20 years diffusionist terminology has lessened, but not disappeared (Bagolini and Biagi 1977), and the role of agricultural colonists is still emphasized (Ammerman and Cavalli-Sforza 1984).

Most current researchers have adopted culture change models within anthropological and ecological frameworks for the explication of the transition to agriculture in Mediterranean Europe. They attempt to identify the people involved in the transition, to describe the process of the transition, and to provide the causes or agents of change. But it is to be observed that in many of these models the principal agents are still colonists; a limited and passive role is given to indigenous mesolithic hunter-gatherers, who are viewed as those affected by change rather than as the principal instigators of and participants in change.

Such a view was presented by Ruth Whitehouse (1968, 1971) who investigated mesolithic and early neolithic cave sites along the Apulian coast in southeast Italy (see Figure 1). Whitehouse modeled the transition to agriculture in a series of stages beginning with mesolithic hunter-gatherer-shellfish collectors. In the second stage domestic sheep and pottery appear within the deposits of these sites along with mesolithic technology and the remains of wild fauna and shellfish. In the third stage evidence appears for the full neolithic economy, including cultigens and domestic animals in addition to sheep. Technology in this third stage was characteristic of traditional neolithic blade industries with transverse arrowheads and obsidian.

Whitehouse suggests that mesolithic hunter-gatherers in southern Italy gained access to neolithic goods and resources through contact with neolithic colonists, probably located in or near the Tavoliere Plain. At first the indigenous hunter-gatherers adopted only sheep

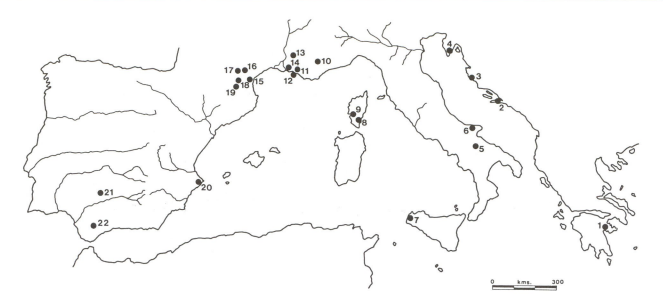

Figure 1. Sites discussed in the text. (1) Franchthi cave; (2) Gudnja Pecina; (3) Skarin Samograd; (4) Medulin Vizula; (5) Rendina; (6) Coppa Nevigata; (7) Grotta Uzzo; (8) Currachiaghiu; (9) Basi; (10) Fontbregoua; (11) Cap Ragnon; (12) Ile Riou; (13) Gramari; (14) Chateauneuf-les-Martigues; (15) Le Correge; (16) Balma Abeurador; (17) Grotte Gazel; (18) Abri de Dourgne; (19) Balma Margineda; (20) Cueva de les Mallaetes; (21) Cueva Chica de Santiago; and (22) Dehesilla.

and pottery. This was soon followed by the formation of a full neolithic economy. Whitehouse views these mesolithic hunter-gatherers as having been under severe pressure to adopt agriculture because of the expanding population of agriculturalists in southeast Italy (Whitehouse 1987).

Whitehouse capably interprets the archaeological record to provide a strong case for her model. Alternatively, Ammerman and Cavalli-Sforza (1984) and others (e.g., Chapman and Muller 1990) suggest that the many cave sites containing ceramics and sheep along with wild game and mesolithic technology in southern Italy and elsewhere are found away from good arable land and were used as special function sites, primarily by pastoralists with flocks of sheep and by groups of neolithic hunters. The associated lithic assemblages thus reflect the specialized nature of these sites.

Ammerman and Cavalli-Sforza (1984) provide a strong theoretical basis for explaining the transition to agriculture in Europe. Their model, influenced by the apparent similarity of the spread of farming and the spread of diseases, involves the initial development of agriculture in Asia Minor followed by its rapid diffusion resulting from differential population growth between farmers and hunter-gatherers, especially in frontier zones (see Dennell 1990). Rapidly growing farming populations expanded the neolithic frontier, disrupting and over-exploiting wild resources and

overwhelming mesolithic hunter-gatherers. Thus the hunter-gatherers of the region were either displaced or became agriculturalists themselves. The role of mesolithic populations in the transition to agriculture in this model is negligible at best.

Ammerman and Cavalli-Sforza's model has received support recently from Sokal et al. (1991) who attempt to provide genetic evidence for the spread of agriculture into Europe through demic diffusion, including the Mediterranean region. While their intensive collection of genetic data is commendable, their argument is weakened by a research design requiring complex multivariate manipulation of the data. More problematical is their failure to take into account the tremendous amount of gene flow that has followed similar routes in historic times (e.g., Greek colonists) and most probably in prehistoric times before and after that associated with the transition to agriculture. This is a situation where results from new but untested "scientific" approaches by those naive of the archaeological record should be carefully scrutinized in light of current archaeological understanding.

Jim Lewthwaite (1989) presents a model which, like that of Whitehouse, identifies sheep as appearing prior to other animal and plant domesticates, but only in the northwest Mediterranean. He suggests that fully developed agricultural communities were present in southern Italy and on the Tyrrhenian Islands of Corsica and Sardinia. From these islands sheep and

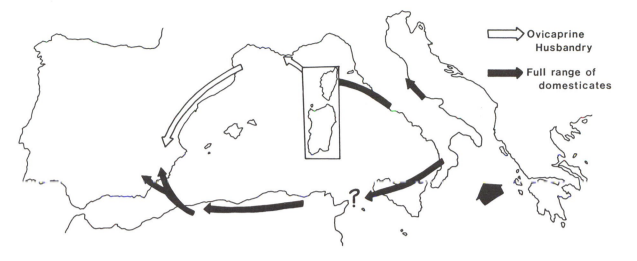

Figure 2. J. Lewthwaite's Island Filter Model

pottery arrived on the mainland in southern France and northern Spain. Meanwhile, fully neolithic settlements spread along the Maghreb of North Africa and arrived in southern Spain (Figure 2). Lewthwaite suggests that the Tyrrhenian Islands acted as a filter, permitting only selected aspects of the agricultural system to be initially transferred to the mainland.

Zvelebil and Rowley-Conwy's (1984) availability model has been applied to the Mediterranean region by Royston Clark (1990), but modified to include a risk reducing strategy developed by Halstead and O'Shea (1982). Zvelebil and Rowley-Conwy's model describes frontier hunter-gatherer/neolithic farmer interaction, and the resulting spatial dynamics. They suggest that before the formation of a contact area between foragers and farmers there is an availability zone where hunter-gatherers are able to gain access to neolithic goods and resources. Later the availability zone develops into a contact zone where increased competition for resources causes farming to be substituted for hunting and gathering. In the last phase the farming system is consolidated into full-scale agriculture. In attempting to use this model for the Adige valley in northern Italy, Royston Clark suggests that groups were exchanging food during times of surplus for craft items and exotic goods that acted as tokens to be exchanged again when food shortages occurred. This created social prestige for some individuals who were then capable of allowing participation in full scale agriculture. The blending of theories directed toward very different social systems has to be severely questioned in light of such little archaeological support.

An alternative to the colonist models was first strongly presented in David Clarke's (1976) provocative essay, "The economic basis of mesolithic Europe,"

which used recent ethnological and archaeological advances in hunter-gatherer research to argue that European prehistorians had greatly exaggerated the significance of wild game and other animal foods, and seriously underestimated the importance of plant foods. Clarke carefully reviewed the biases of the archaeological record including differential preservation of the food remains and of the technology required by the two food categories, and of the misinterpretation of the use of the technology that has been recovered.

Clark suggested that prior to the transition to agriculture in Europe, late paleolithic and mesolithic hunter-gatherers were transhumant and primarily focused on reliable plant resources in the region. With changing landscapes and climates, large migratory herbivores became less reliable as a food resource, leading toward the eventual collapse of the transhumant system in favor of marine resource exploitation during summer. Plant foods became increasingly important; with increasing sedentism there was greater ability, and perhaps greater need, to control and manipulate such resources. During this time of intensive plant use, and plant and animal manipulation, the neolithic domesticates became available. Because of their advantages in productivity and storage, neolithic domesticates rapidly replaced wild resources, and eventually reduced the role of marine resources as well.

Chapman and Muller (1990) adopted Clark's model for the eastern Adriatic, claiming that their site data support Clarke's agriculture replacement model. They emphasize that there is no need to consider a pastoral hunter-gatherer phase prior to the formation of open-air agricultural settlements.

John Evans (1987) follows a similar line of reasoning for the southeast Adriatic and for southern Italy. He argues for an introduction and rapid spread of neolithic domesticates and technology leading toward the formation of fully developed agricultural communities at about 7000 B.P.

David Geddes has presented a model based on finds of domestic sheep in late mesolithic levels at Grotte Gazel, Abri de Dourgne, Chateauneuf-les-Martigues, and Gramari in southern France and also reported at Cueva de les Mallaetes (Geddes 1985). According to Geddes, these mesolithic levels do not contain any other evidence of neolithic influence or material, including pottery. Younger levels at these sites produced other domesticates, cultigens, and pottery. The technology associated with the domestic sheep is mesolithic and the stratigraphic sequences display a clear transition toward traditional neolithic technology. For example, among the artifacts associated with this transition in the Aude valley are microlithic geometrics that gradually change to a triangular form with backing followed by a triangular form with invasive or flat retouch (Geddes et al. 1989). Geddes's model of mesolithic pastoralism has been challenged by Iain Davidson (1989) who suggests that the association of domestic sheep in mesolithic levels indicates only that domestic sheep had escaped from neolithic farmers and been hunted along with wild game.

Halstead (1989) and Perlés (1988) have presented models that are specific to the Aegean and Greece. It appears that two settlement types are emerging, permanent villages similar to the tells of the Near East, and seasonally occupied sites such as Franchthi cave. Importantly, one observes major differences in technology and economy between the site types. Franchthi cave, for example, displays a strong lithic continuity from the mesolithic through the neolithic.

Discussion

Research on the transition to agriculture in the Mediterranean Basin continues the very old debate of colonization by foreign folks vs. acculturation by native mesolithic inhabitants of the region. What reasons are there for rejecting indigenous culture change among mesolithic hunter-gatherers? There are five primary points to address: (1) domesticated plants and animals associated with the neolithic come from the Near East; (2) the time available between the appearance of agricultural products in the Mediterranean Basin and the formation of fully sedentary agricultural villages was too short to result from acculturation; (3) there was simultaneous introduction of pottery, polished stone tools, cultigens, and domesticates; (4) the earliest fully agricultural communities appear along or near the coast rather than in the

interior; and (5) there has been little evidence for intensive wild plant use in the mesolithic as a precursor to agriculture.

These arguments are now insufficient to reject the significant role of the indigenous hunter-gatherers. It is clear that all early domesticates are from Asia Minor. With regard to the early domesticates, the fact that their origins are to the east is a geographic point and has little bearing on the hypothesis of indigenous culture change. It indicates only that mesolithic populations in the eastern Mediterranean probably had access to domesticates earlier than populations in the western Mediterranean. The most controversial of possible domesticates in Europe has been sheep which was suggested as being locally domesticated from the French mouflon of France or Corsica (Dennell 1983). This has been demonstrated to be false by paleontological, anatomical and genetic research, and it is recognized now that the Corsico-Sardinian mouflon is a feral population of early domestic sheep (Lauvergne 1977; Nguyen and Bunch 1980; Poplin 1979; Vigne 1983).

Results from Franchthi Cave indicate that boats were navigating the Mediterranean as early as the late Upper Paleolithic (Jacobsen 1976). Domesticated animals and cultigens may have been transported not only by neolithic traders, but also by mesolithic seafarers, who could have extended the availability of agricultural products well beyond Anatolia and the Levant. In this regard, the excellent review by Cherry (1990) of the early settlement of Mediterranean islands must be considered. It is evident that many islands such as Sardinia and Corsica were visited prior to the arrival of full-scale agriculture. What is important here is that the evidence to date suggests that the Mediterranean islands, even the larger ones, could not sustain reproductively viable populations of hunter-gatherers.

It is no longer valid to claim that there is insufficient time for hunter-gatherers to make the transition to farming. Domestic sheep are being recovered in France in mesolithic cave sites with dates falling between 7800 and 7300 b.p. These sites include: Chateauneuf C8 and C7 at 7830±170 (Ly-438), 7525±100 (MC-unpub.), 7270±220 b.p. (Ly-448) (Schvorer et al. 1977:34); Gramari at 8000±190 (Gif-753) and 7740±190 b.p. (Gif-752) which are the more accepted later dates (Poulain in Dumas et al. 1971, Dumas et al. 1979); Grotte Gazel F6-Porch at 7880±100 (GrN-6704); and Abri de Dourgne C7 at 6850±100 b.p. (MC-1107) (Geddes 1985). Meanwhile reassessment of the dates of fully agricultural sites in France and Spain has invalidated those sites dated prior to 7000 b.p. These include such French sites as Cap Ragnon and Ile Riou which were dated with marine shell. The southern Spanish sites of

Cueva Chica de Santiago and Dehesilla have been re-dated around 6200 B.P. (6160±100 [UGRA-254] and 6260±100 B.P. [UGRA-259] respectively) (Lewthwaite 1989). The very early Corsican neolithic sites of Basi and Currachiaghiu have been questioned because of the late Cardial style of the associated ceramics (Guilaine 1980). These reassessments mean that domestic sheep were in the western Mediterranean a full millennium before the appearance of sites with a fully neolithic economy. A 7000 B.P. date for the fully neolithic in Spain and France fits well with the pattern now emerging in southern Italy and the Adriatic. As more full neolithic economy sites in southern Italy are dated, support grows for their age between 7500 and 7200 B.P. (Whitehouse 1987). This is further supported by evidence for fully neolithic sites from the Dalmatian coast around 7200 B.P. (Chapman and Muller 1990) and from the southeastern Adriatic and Greece by 7500 B.P. (Whitehouse 1987:366).

Contrary to Chapman and Muller (1990), virtually all regions of Mediterranean Europe witnessed the appearance of domestic sheep and ceramics prior to the advent of sites with fully neolithic economy. Even in the eastern Adriatic, caprines are the only identified domestic species at the early sites of Skarin Samograd and Medulin-Vizula. The idenfication of domestic pig at Gudnja Pecina may instead be wild boar.

The first appearance of neolithic sites along on the coasts of Mediterranean Europe has been the principal basis for the colonist model. However, Phillips (1987) has recently noted the association between coastal lowlands and the appearance of the full neolithic sites, and further that the best arable lands are also located along the coast. It is therefore not surprising that the earliest use of cultigens occurs at sites located on good arable lands, near where resources first became available, and where there was the greatest complement of wild and marine food resources to support permanent settlements.

Now that evidence for plant use is being sought systematically, there is greater support for Clarke's hypothesis. More than fifteen species of wild plants have been recovered from mesolithic levels of Franchthi cave in Greece (Hansen 1978), five species have been identified from the Epipaleolithic levels of Grotta dell'Uzzo in Sicily (Hopf 1991), and nineteen species have been identified from mesolithic levels of cave sites in southern France and northern Spain. These principally include the mesolithic sites of Balma Abeurador, Balma Margineda and Fontbregoua where there are nuts, pulses, fruits, tubers, greens, and wild cereals (Hopf 1991; Vaquer et al. 1986).

These data are not fully in support of Clarke's model, however. It appears that his model falters in many ways. Although plant foods are demonstrably important during the mesolithic in Mediterranean Europe, there is still no evidence for intensive collecting, processing, or storing of plant foods that would suggest extended use over winter or over the dry late summer months. Clarke's error in this regard may relate to the plant species once thought to have been available to Mediterranean hunter-gatherers. For example, chestnut was not available in the Mediterranean region until the Classical Period (Zohary and Hopf 1988). There is also no evidence for intensive use of acorns. The California model of intensive acorn (or other nut) gathering is not valid for Mediterranean Europe (Lewthwaite 1983; Phillips 1975).

This conclusion is supported by growing evidence for the continuation of a transhumant settlement system in the late Upper Paleolithic and mesolithic throughout much of the region (Barker 1981; Donahue 1988; Geddes 1985). Frequent foraging for plant foods, rather than intensive long-term storage strategies, would reinforce the transhumant pattern; not only would wild game be in the uplands during the summer and autumn, but so would most plant foods. As Hopf (1991) notes, such movement would allow plant foods to be available to human groups throughout the year.

This transhumant settlement system and absence of long-term plant food storage may explain in part the temporal differences in the adoption of agricultural resources. Sheep can be easily assimilated within a transhumant hunter-gatherer subsistence-settlement system; their maintenance is very much a risk-reducing food storage strategy. Equally important, small-scale pastoralism would unlikely conflict with the exploitation of wild resources; the shepherding of sheep could be performed by men too old to hunt and by adolescent children. A major reorganization of society is not required in order to adopt agriculture, as suggested by Royston Clark (1990).

Why should we re-think the agricultural transition? Domestic sheep appear by 7800 B.P. in the western Mediterranean. Fully agricultural sites are measurably later in date than pastoral sites in the same region. Wild plants, in common use throughout the mesolithic, continued to be exploited throughout much of the neolithic. The transhumant settlement system continued throughout the mesolithic and neolithic, and mesolithic caves and open-air sites continued in use in the neolithic. The mesolithic technology, whether Castelnovian, Tardonesian, Romanellian, or another regional variant, continued into the early neolithic, albeit with gradual changes. Ceramics are found in sites with distinctly mesolithic stone artifacts. It appears that mesolithic groups were adopting ceramics at an early date in some regions and not in others; the ceramics may or may not be associated with domestic sheep and vice versa. However, these regional variations in the transition to agriculture are to be expected given differences in environment, economy, and society.

Research continues to identify early neolithic open-air settlements containing the full assemblage of the neolithic economy, such as La Correge in France (Geddes 1985) and Rendina in Italy (Whitehouse 1987), which do not date earlier than the pastoral sites. The continued intensive use of caves throughout the neolithic, even when the fully neolithic economy is in place, requires explanation. Often these were mesolithic occupation sites, and the continuity in occupation provides additional support to an argument for indigenous culture change. Finally, one observes that transhumant settlement systems, with seasonally-occupied special function sites, continued throughout the neolithic even after the formation of sites with the full neolithic economy through the western Mediterranean (Barker 1981; Geddes 1983; Guilaine et al. 1982).

Colonist models and their supportive data are being challenged. Many early dates for fully agricultural sites are no longer accepted, e.g., Coppa Nevigata. As recently suggested by David Anthony (1990), if a migration theory is to be considered for the Mediterranean, a model consisting of small coastal colonies with continued trade and communication with the homeland is a far more likely scenario than a wave of advance as proposed by Ammerman and Cavalli-Sforza (1984).

Conclusion

While it is relatively easy to claim that neolithic colonists were the source of domestic resources and new technologies throughout the Mediterranean, the accumulating evidence leads one to reject that view. I suggest it is time for archaeologists to consider more seriously the transition to agriculture in terms of the changing adaptations of indigenous hunter-gatherers. We now need to address issues concerning the process of that change, and the causes for the differences observed in that process, among the different regions of the Mediterranean.

References Cited

Ammerman, Albert. J., and L.L. Cavalli-Sforza.
  1984 *The Neolithic Transition and the Genetics of Populations in Europe.* Princeton University Press, Princeton, N.J.
Anthony, David.
  1990 Migration in Archaeology: The Baby and the Bathwater. *American Anthropologist* 92:895-914.
Barker, Grahame
  1981 *Landscape and Society.* Academic Press, New York.
Bernabó Brea, Luigi
  1956 *Gli Scavi nella Caverna delle Arene Candide 2.* Istituto di Studi Liguri, Bordighera.
Bagolini, Bernardino, and Paolo Biagi
  1977 Current Culture History Issues in the Study of the Neolithic of Northern Italy. *Institute of Archaeology Bulletin* 14:143-166.
Chapman, John, and Johannes Muller
  1990 Early Farmers in the Mediterranean Basin: The Dalmatian Evidence. *Antiquity* 64:127-34.
Cherry, John F.
  1990 The First Colonization of the Mediterranean Islands. *Journal of Mediterranean Archaeology* 3:145-221.
Clarke, David
  1976 Mesolithic Europe: The Economic Basis. In *Problems in Economic and Social Archaeology*, edited by G. de G. Sieveking, I.H. Longworth, and K.E. Wilson, pp. 449-81. Duckworth, London.
Clark, Royston
  1990 The Beginnings of Agriculture in Sub-Alpine Italy: Some Theoretical Considerations. In *The Neolithisation of the Alpine Region*, Monografie di "Natura Bresciana" 13, edited by Paolo Biagi, pp. 123-137.
Davidson, Iain
  1989 Escaped Domestic Animals and the Introduction of Agriculture to Spain. In *The Walking Larder*, edited by David Harris and G. Hillman, pp. 59-71. Unwin Hyman, London.

Dennell, Robin

    1983　*European Economic Prehistory.* Academic Press, New York.

    1990　The Hunter-gatherer/Agricultural Frontier in Prehistoric Europe. In *The Archaeology of Frontiers and Boundaries,* edited by S. Green and S. Perlman, pp. 113-139. Academic Press, Orlando.

Donahue, Randolph E.

    1988　Settlement, Seasonality and Site Function in the Italian Final Epigravettian. Paper presented at the 54th Meeting of the Society for American Archaeology. Phoenix.

Dumas, C., M. Livache, Jean-Claude Miskovsky, and T. Poulain

    1971　Le camp Mésolithique de Gramari à Méthamis (Vaucluse). *Gallia Préhistoire* 14:47-137.

Dumas, C., M. Livache, Jean-Claude Miskovsky, and M. Paccard

    1979　A propos de datations ^{14}C de Gramari (Méthamis). In *La Fin des Temps Glaciaires en Europe,* edited by D. de Sonneville-Bordes, pp. 43-45. C.N.R.S., Paris.

Evans, John

    1987　The Development of Neolithic Communities in the Central Mediterranean: Western Greece to Malta. In *Prémieres Communautes Paysannes en Méditerranée Occidentale,* edited by Jean Guilaine, Jean Courtin, Jean-Louis Roudil, and Jean-Louis Vernet, pp. 321-327. C.N.R.S., Paris.

Geddes, David

    1983　Neolithic Transhumance in the Mediterranean Pyrenees. *World Archaeology* 15: 51-66.

    1985　Mesolithic Domestic Sheep in West Mediterranean Europe. *Journal of Archaeological Science* 12: 25-48

Geddes, David, Jean Guilaine, Jacques Coularou, Oliver Le Gall, and Michel Martzluff

    1989　Postglacial Environments, Settlement and Subsistence in the Pyrenees: the Balma Margineda, Andorra. In *The Mesolithic in Europe: Papers Presented at the Third International Symposium, Edinburgh 1985,* edited by Clive Bonsall, pp. 561-71. John Donald, Edinburgh.

Guilaine, Jean

    1980　Problémes actuels du Néolithiques Ancien en Méditerranée Occidentale. In *Interaction and Acculturation in the Mediterranean,* edited by J. Best and N. de Vries, pp. 13-15. Gruner, Amsterdam.

Guilaine, Jean, Michel Barbaza, David Geddes, Jean-Louis Vernet, Miguel Llongueras, and Maria Hopf

    1982　Prehistoric Human Adaptations in Catalonia (Spain). *Journal of Field Archaeology* 9: 407-416.

Halstead, Paul

    1989　Like Rising Damp? An Ecological Approach to the Spread of Farming in Southeast and Central Europe. In *The Beginnings of Agriculture,* edited by A. Milles, D. Williams, and N. Gardner, pp. 23-53. British Archaeological Reports, International Series, 496. BAR, Oxford.

Halstead, Paul, and John O'Shea

    1982　A Friend in Need is a Friend Indeed: Social Storage and the Origins of Social Ranking. In *Ranking, Resource and Exchange,* edited by Colin Renfrew and Stephen Shennan, pp. 92-99. Cambridge University Press, Cambridge.

Hansen, Julie Marie

    1980　*The Palaeoethnobotany of Franchthi Cave, Greece.* Ph.D. dissertation, University of Minnesota. University Microfilms, Ann Arbor, Michigan.

Hopf, Maria

    1991　South and Southwest Europe. In *Progress in Old World Palaeoethnobotany,* edited by W. Van Zeist, R. Wasylikowa, and K. Behre, pp. 241-276. Balkema, Rotterdam.

Jacobsen, Thomas W.

    1976　17,000 Years of Greek Prehistory. *Scientific American* 234(6): 76-87.

Lauvergne, J.

    1977　Utilisation des marquers génétiques pour l'étude de l'origine et de l'évolution du mouton domestique. *Ethnozootechnie* 21: 17-24.

Laviosa Zambotti, Pia

    1943　*Le Piú Antiche Culture Agricole Europée.* Giuseppe Principato, Milan.

Lewthwaite, James

    1983　Acorns for the Ancestors: The Prehistoric Exploitation of Woodland in the West Mediterranean. In *Archaeological Aspects of Woodland Ecology,* edited by M. Bell and S. Limbrey, pp. 217-230. British Archaeological Reports, International Series, 146. BAR, Oxford.

1989 Isolating the Residuals: The Mesolithic Basis of Man-Animal Relationships on the Mediterranean Islands. In *The Mesolithic in Europe: Papers Presented at the Third International Symposium, Edinburgh 1985*, edited by Clive Bonsall, pp. 541-55. John Donald, Edinburgh.

Nguyen, T., and T. Bunch
1980 Blood Groups and Evolutionary Relationships among Domestic Sheep (*Ovis aries*), Domestic Goat (*Capra hircus*), Aoudad (*Ammortragus lervia*) and European Mouflon (*Ovis musimon*). *Annales de Génétique et de la Sélection Animale* 12: 169-180.

Perlés, Catherine
1988 New Ways with an Old Problem: Chipped Stone Assemblages as an Index of Cultural Discontinuity in Early Greek Prehistory. In *Problems in Greek Prehistory*, edited by E.B. French and K.A. Wardle, pp. 477-488.

Phillips, Patricia
1975 *Early Farmers of West Mediterranean Europe*. Hutchinson, London.
1987 The Development of Agriculture and Peasant Societies in the West Mediterranean. In *Prémieres Communautes Paysannes en Méditerranée Occidentale*, edited by J. Guilaine, J. Courtin, J.-L. Roudil and J.-L. Vernet, pp. 275-78. C.N.R.S., Paris.

Poplin, F.
1979 Origine du mouflon de Corse dans une nouvelle perspective Paléontologique: par Marronnage. *Annales de Génétique et de la Sélection Animale* 11: 133-43.

Schvorer, M., C. Bordier, J. Evin, and G. Delibrias
1979 Chronologie absolué de la fin des temps glaciaires en Europe. Recensement et présentation des datations se rapportant à des sites Français. In *La Fin des Temps Glaciaires en Europe*, edited by D. de Sonneville-Bordes, pp. 21-41. C.N.R.S., Paris.

Sokal, Robert R., Neal L. Oden, and Chester Wilson
1991 Genetic Evidence for the Spread of Agriculture in Europe by Demic Diffusion. *Nature* 351: 143-45.

Vaquer, Jean, David Geddes, Michel Barbaza, and Jean Erroux
1986 Mesolithic Plant Exploitation at the Balma Abeurador (France). *Oxford Journal of Archaeology* 5: 1-18.

Vigne, J-D.
1983 *Les Mammiféres Terrestres Nonvolants du Postglaciare en Corse et leurs Rapports avec l'Homme*. Unpublished Ph.D. dissertation, Université de Paris VI.

Whitehouse, Ruth
1968 The Early Neolithic Sequence of Southern Italy. *Antiquity* 42: 188-93.
1971 The Last Hunter-gatherers in Southern Italy. *World Archaeology* 2: 239-54.
1987 The First Farmers in the Adriatic and their Position in the Neolithic of the Mediterranean. In *Prémieres Communautes Paysannes en Méditerranée Occidentale*, edited by J. Guilaine, J. Courtin, J.-L. Roudil, and J.-L. Vernet, pp. 357-366. C.N.R.S., Paris.

Zohary, Daniel, and Maria Hopf
1988 *Domestication of Plants in the Old World*. Clarendon, Oxford.

Zvelebil, Marek, and Peter Rowley-Conwy
1984 Transition to Farming in Northern Europe: A Hunter-gatherer Perspective. *Norwegian Archaeological Review* 17(2): 104-125.

7

The Introduction of Agriculture
to the Western North European Plain

Lawrence H. Keeley
University of Illinois at Chicago

The first appearance of agricultural economies has long been a focus of archaeological interest in many parts of the world. During the 60s and 70s, such interest concentrated almost exclusively on those regions where agriculture first developed independently from a hunting and gathering base, i.e. the "pristine agricultural hearths" of Southwest Asia, Mesoamerica, highland South America, and, more problematically, the tropics of Southeast Asia, Africa and Amazonia. This attention to such pristine hearths was accompanied, and to some degree generated, by a flurry of theories concerned with the causes and processes that led hunter-gatherers to invent agriculture. Archaeologists working in regions outside of these obvious hearths either glumly accepted their theoretical irrelevance or scoured the archaeological record of their proprietary region for any evidence, however slight, that some independent steps toward agriculture were taken there.

For a variety of reasons, including the focus of the New Archaeology on social evolution, moral reactions to colonialism, and regional chauvinism, the processes of migration and diffusion became disregarded and disapproved (Anthony 1990; Otte and Keeley 1990). By the later part of the 1980s, however, better genetic, archaeozoological, and palaeobotanical data made hypotheses of independent domestication difficult to sustain in most areas outside of the already-recognized hearths (e.g., Geddes 1985). Recent attention to such topics as farmer-forager interactions and

"neolithization" has revived interest in the processes involved in the diffusion of domesticated plants and animals to prehistoric foragers. However, the migration of people remains a neglected theoretical area and one that is largely taboo for the interpretation of prehistoric change, including the appearance of agriculture.

The introduction of agriculture into northwestern Europe, then, is inherently interesting since it involved both an indisputable migration of farmers, and the diffusion of certain Neolithic features, specifically domesticated livestock and ceramics, to indigenous hunter-gatherers. The most striking feature of this story is that these two processes appear to have operated quite independently of one another. As I will argue, the neolithic traits which appeared among the indigenous Late Mesolithic hunter-gatherers were not borrowed or learned from the colonizing farmers.

Study Region

The area considered in this paper is the western extension of the North European Plain. The area includes all of Lower and Middle Belgium, north and west of the Meuse, and the immediately adjacent area of southern Holland. As with the other parts of the Plain, the loess deposits to the south grade into the Coversands of the north. During the period of interest from 7000 to 5500 B.P. (uncorrected), the primary vegetation cover on the loess was Oak-Elm-Linden forest.

The Linearbandkeramik (LBK) Colonization

Perhaps the most easily and widely recognized prehistoric culture in Europe is the Linear-bandkeramik (LBK, also known as Linear Pottery and Rubané) which has uncalibrated radiocarbon dates ranging between 6700 and 5900 B.P. LBK is distributed from northern France to the western Soviet Union. It is easily identified from its distinctive ceramics, house forms, and lithics which are remarkably homogeneous over great distances.

LBK Material Culture

LBK habitations were long houses 6 to 7 m wide, varying in length from 12 to 45 m (Figure 1). The walls were constructed of wattle and daub and the roof supported by a number of sets of triple posts (tierces). This rather invariant house format may be found throughout the LBK distribution in Europe. The variation in house length is, for the most part, modular. The three modules are head, center, and foot. Thus, the smallest houses include only the central module, the medium type contains the central and head modules, while the largest types have all three modules. The significance of these size differences is unclear since they show no chronological correlations. In some sites, even entire regions, nearly all houses are of the largest type.

The LBK is named for its unpainted fineware which was almost invariably decorated with impressed or inscribed bands (Figure 2). Fineware typically received a reduced firing which gives it a black or gray color. The undecorated coarseware was thicker walled and fired so as to have a black interior but orange- or yellow-buff exterior. The most common tempers are grog and grit. Almost all vessel shapes, both coarse and fine, are based on the sphere or hemisphere; flat-bottomed forms are very rare (except in the Most Ancient LBK, restricted to Central Europe). The Most Ancient and, to a slightly lesser degree, the Early LBK ceramics are stylistically homogeneous over their entire distribution. Regional styles can only be observed in the Late LBK.

At Early and Late LBK sites in the Low Countries, western Germany, and northern France, in addition to the standard LBK ceramics, one commonly finds a few vessels of the enigmatic Limburg Ceramic (LC). LC is very different from LBK wares in temper (e.g., bone temper), firing, form and decoration; it is very unlikely to be the product of LBK potters. LC is mentioned here because Modderman (1981) has argued that it was manufactured by Late Mesolithic groups. The problem with this interpretation is that, with a few minor exceptions where isolated LC sherds were found beyond the

loess region, LC is always found as a minority ware in LBK contexts. LC has never been recovered from *any* Mesolithic site. It has been compared with early ceramics from the Eastern Baltic (Constantin 1985) and southwest France (Van Berg and Cahen in press), but its local origin and the reasons for its persistent association with LBK remain an irritating mystery.

The lithic industries of the Belgian LBK were produced by classic blade reduction technique from unidirectional cores which yielded long, broad blades quite unlike the narrow bladelets of the late Mesolithic. The most common and ubiquitous tool types are endscrapers, sickle elements, exhausted blade cores used as hammerstones to resurface grinding stones, and the distinctive Danubian point (Figure 3). There are a number of other tool types which are more infrequent and less uniformly distributed: denticulates, piercers, pièce esquillée, frits (a peculiar steep-edged blade with a distinctive usewear pattern), orange quarters (essentially, a retouched frit) and notched pieces. Burins are extremely rare and the few examples found seem to have been accidents. Some regional variations in lithic assemblages do occur but they seem primarily explained by variation in the quality and availability of flint. Milisauskas (1978; Milisauskas and Kruk 1989) has repeatedly noted that projectile points are much more common in the Western than in the Eastern LBK assemblages. There is, however, no evidence that hunting was more significant in the western region. Indeed, hunting seems to have been rather unimportant to all LBK groups (see below).

The LBK is also characterized by two distinctive groundstone implements: saddle querns and "shoe-last" adzes (Figure 3). These adzes were not just wood-working tools. The data from the Nitra LBK cemetery in Czechoslovakia implies they were male status symbols since they were preferentially interred with older males (Sherratt 1976). The recent find, at Talheim in Germany, of a mass grave with the bodies of 35 men, women and children killed by blows to the head with shoe-last adzes, gives unequivocal evidence that the adze was also a weapon (Wahl and König 1987).

LBK material culture shows very limited stylistic variability across its huge geographic distribution. I had no difficulty recognizing as LBK, ceramics, adzes and querns in Soviet Moldavia although all my previous experience had been with westernmost LBK. Such uniformity attests to a rapid spread, or an almost pathological conventionality, or both.

LBK Subsistence

The LBK subsistence economy was almost entirely agricultural. Staple crops included einkorn and emmer

Rubané linéaire recent   Rubané linéaire ancien

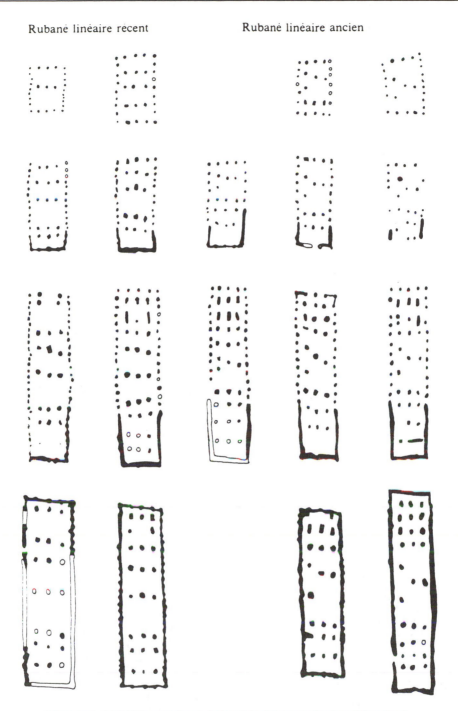

Figure 1. LBK Houseplans (after Modderman's classification).

wheat, (spelt, bread, and club wheat are unknown in the Benelux and very rare to the east), barley (which is very rare in the Benelux), peas, lentils, and flax. Except for hazelnuts, wild plant foods seem to have been unimportant since most of the wild plants recovered at LBK sites would have grown as weeds in the grain fields. The acid soils of the loess seldom preserve bone, but in cases where there is such preservation some consistent patterns of animal exploitation emerge: (1) 80% to 95% of the mammal bone found at Early Neolithic sites were from domesticates; (2) usually over half of such remains were of domestic cattle, ovicaprids being a distant second, and pigs were usually rare (Bogucki 1988; Milisauskas and Kruk 1989). Many authors (e.g., Milisauskas and Kruk 1989) have commented on the peculiar LBK lack of enthu-

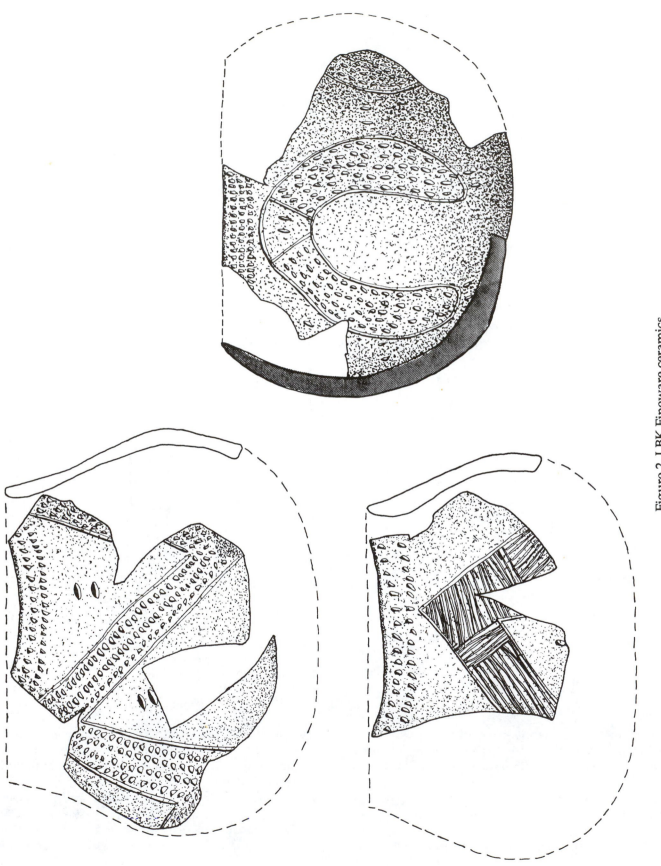

Figure 2. LBK Fineware ceramics.

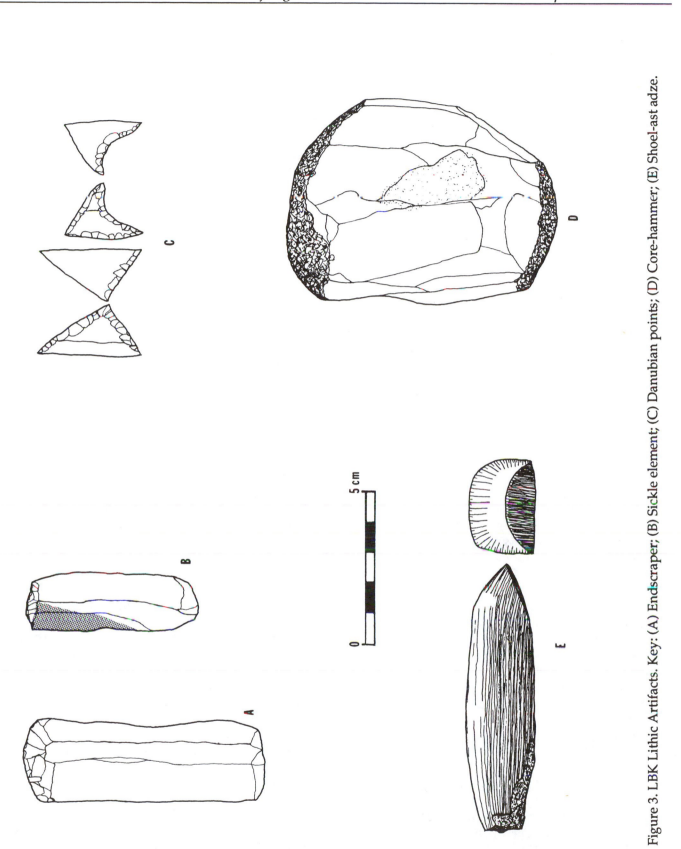

Figure 3. LBK Lithic Artifacts. Key: (A) Endscraper; (B) Sickle element; (C) Danubian points; (D) Core-hammer; (E) Shoel-ast adze.

siasm for swine, since the latter are so well adapted to and productive in the deciduous forest. There is also evidence from Place St. Lambert, where bone preservation was particularly good, that fish were taken in quantity by LBK folk, in this instance, during the winter (Desse 1984). Given that these were mostly of minnow size, they were probably taken with nets or traps.

The picture then is of an almost entirely agricultural economy, whose staples were wheat and beef, with little or no use of wild foods, except hazelnuts and possibly fish. The traditional picture of the LBK agricultural system once was that it was of the long-fallow, slash-and-burn type with villages abandoned every generation. This view is no longer accepted by the majority of prehistorians (e.g., Bakels, Barker, Bogucki, Gregg, Milisauskas, Modderman, Rowley-Conwy, Whittle, etc.), who persuasively argue that with rotation of grains, legumes, and pasture, with manuring, etc., yields could have been sustained for very long periods. Indeed, the size and density of houses at some LBK sites strongly implies that at least some LBK villages (e.g., Elsloo, Sittard, Köln-Lindenthal) were occupied essentially continuously over many generations.

Gregg (1988) has proposed a model of LBK land use which requires only minimal forest clearance. Our palynological data from Belgium do not agree with this model since we have evidence that forest clearance was both rapid and extensive at pioneer LBK villages, with a concomitant increase in the pollen of cereals and pasture weeds (Keeley and Cahen 1989; Heim 1983). There is some palynological evidence, after clearance, that a considerable amount of hazel and some linden remained. Even at sites where we have evidence of several phases of occupation, such as Oleye, there is no palynological evidence suggesting forest regeneration. It seems, then, that LBK land use involved rapidly clearing an area of the Atlantic forest of most of its trees, except for hazel bushes and a few lindens, to create fields for crops and pastures for livestock. All of the above evidence and arguments imply that such drastic ecological changes, involving the substitution of domestic plants and animals for those native to the Atlantic forest, were then maintained over many generations.

LBK Settlement Pattern and Social Units

LBK settlement preferences have been extensively studied across the area of their distribution (see Whittle 1987 for a recent review). Given the great homogeneity in material culture and subsistence, it is not surprising to find a similar uniformity in settlement patterns. There is almost exclusive preference for loess or loess-derived soils. The valleys of major rivers were avoided in preference for second- and third-order streams. Settlements are located on gentle slopes just above valley bottoms; interfluves, valley bottoms, and plateaus were avoided. Palynological evidence from Darion indicates pastures were downslope and grain fields upslope from the village (Hiem 1983). As no one has discovered any difference in settlement preference between the earliest and latest LBK sites in any region, these "rules" must be considered as "traditional" and a part of the larger, correlated complex of traits that characterize LBK culture.

LBK settlements occur in small clusters of 10 to 20 sites. Bakels (1982) and Lüning (1982) argue that these clusters represent some type of social unit but the economic role of, if any, and sociopolitical nature of such units is obscure. In our work on several LBK sites in the Upper Geer cluster, we found evidence for part-time, village-level specializations in lithic blade and ceramic production (Keeley and Cahen 1989), as well as manufacture of some as yet unspecified types of hide and wood products (Keeley and Cahen 1990). None of these specializations can be explained by reference to differential access to necessary raw materials, since the villages were separated by no more than 3 km and the flints, clays, hides, and woods would have been equally accessible to all.

There is no evidence that these specializations developed gradually. They appear to have been present at the inception of each settlement, implying they were part of the cultural "baggage" carried by the LBK folk who established the Upper Geer settlement cell. We have suggested that these specializations supported military alliances between a series of small, vulnerable settlements (Keeley and Cahen 1990).

There is also evidence in the homogeneity of raw materials for adzes found at the separate villages in the Upper Geer cluster, and its sharp contrast with the neighboring Yerne cluster, that "pooling" and centralized redistribution of these products was a feature of the socioeconomic organization of clusters (Keeley and Cahen 1989). The labor for the construction and defense of the fortifications around these villages would also have required the cooperation of several communities (Keeley and Cahen 1989). Lüning (1982) argues that the construction of the LBK "enclosures" on the Aldenhovener Platte was also the result of cooperation among several villages. Thus, the primary socioeconomic unit of the LBK appears to have been an intentionally economically interdependent group of allied villages, rather than autarchic, independent households or villages—at least in the our region of the LBK distribution.

Discussion

LBK then is an extensive, coordinated complex of stylistic and economic traits which are invariably found associated with one another in thousands of sites distributed over a huge area of the Middle Danube Basin and the loess lands of the North European Plain.

None of these characteristics in lithic technology, ceramics, house forms, settlement location, mixed-farming subsistence, and so forth, can be found in the Mesolithic cultures that preceded the LBK within its distribution area. It is scarcely credible that the spread of LBK culture could be the result of acculturation or diffusion. This would imply a stupefying, wholesale, collective forgetfulness about their "old" Mesolithic lifestyle and a complete and instantaneous acceptance of a totally different and "new" LBK way of life, even to the finest details of ceramic decoration and lithic technology, with no exceptions or errors. Most prehistorians accept that the spread of LBK is the result of colonization by new folks.

Documentation of this migration, however, is not without its difficulties. The distribution of the different ceramic stages clearly conforms to the expectations of the colonization hypothesis with the Most Ancient LBK restricted to Central Europe and the later stages reaching successively further north, east, and west. But radiocarbon cannot distinguish between these stages, nor are the earliest dates from Central Europe significantly older than the oldest dates from the Low Countries; only the dates from northern France are significantly younger than the rest (Cahen and Gilot 1983). Perhaps this indeterminacy is partially explained by the LBK use of wood from climax forest trees whose heartwood may date several hundred years older than their bark. Averaging of dates should have eliminated some of this problem, but did not. The only reasonable conclusion is that the migration from Central to Northwestern Europe was extremely rapid and took only 200-300 years (Cahen and Gilot 1983).

Late Mesolithic Developments

Except for lithic technology, the way of life of Late Mesolithic of our area is rather poorly known because organic remains are not preserved at most sites, materials are often vertically dispersed and mixed by bioturbation, excavations are typically small in scale, and radiocarbon dating seems to yield erratic results, especially from open-air sites (Gilot 1984).

Late Mesolithic Material Culture

In discussing the lithic technology of the Late Mesolithic, I will endeavor to avoid the lengthy description of detailed typological variation which seems endemic to this subject. The Late Mesolithic of our region is characterized by production of thin, regular bladelets by a technique known as Montbani debitage. Various local flints are used but Wommersom quartzite, a chert-like material from only one known source, is a widely-distributed and popular import (Gendel 1982). The type fossils of these industries are "mistletoe" or surfaced-retouched points, trapeze-shaped microliths, and bladelets with Montbani retouch (Figure 4). Gendel (1984) has noted that the distribution of Wommersom quartzite and mistletoe points almost precisely coincide. This techno-complex has been named the Rhine-Meuse-Schelde or RMS.

The dating of late RMS is erratic but accepted dates ranges from 7800 B.P. until about 6000 B.P. (Gob 1984b, De Laet 1982). In other words, the final stages of RMS are contemporary with LBK. Occasional finds of Danubian-style points and, much more rarely, LBK adzes in later RMS sites is also cited as evidence that its final stages were contemporary with LBK (De Laet 1982; Gob 1984b).

The only habitation structure yet recovered for RMS is an elliptical ring of cobbles centered on a slab-lined hearth from Leduc (Gob and Jacques 1985; Gob 1984b). This structure has a ^{14}C date of 6990±90 B.P. and therefore predates the appearance of LBK in the region.

The overwhelming majority of RMS sites have produced no ceramics. In Belgium there are only three sites that have yielded possible associations of RMS lithics and ceramics: Weelde-Paardsdrank 4/5, Oleye (Feature 88094), and Melsele, the most secure association.

At the Weelde sites, small sherds of undiagnostic bone-tempered ceramic were found in the same bioturbated layer as classic late RMS lithics (Huyge and Vermeersch 1982). The primary evidence for the lithic-ceramic association is the similar vertical and horizontal distributions of these two materials. There are two ^{14}C dates of 6990±135 and 5710±80 (Gilot 1984).

At the LBK site of Oleye, finds of Late Mesolithic material in the fill of LBK pits were rather common suggesting that some sort of Mesolithic occupation had occurred (Keeley and Cahen 1989). One pit, however, of a form unusual for the LBK, contained a small number of Mesolithic artifacts (mostly of Wommersom quartzite) and a few very small sherds of bone-tempered ceramic. No datable organic materials were recovered, but various evidence suggests that the pit pre-dates the LBK occupation which began between 6500 and 6300 B.P.

Melsele, near Antwerp, is a site similar to Weelde with classic RMS artifacts and ceramic sherds contained in a bioturbated layer at the top of a low sand dune (Van Roeyen and Van Hoven 1989). At

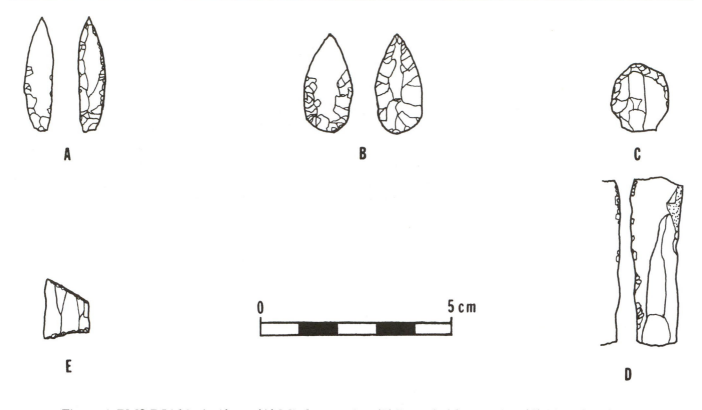

Figure 4. RMS-B Lithic Artifacts. (A) Mistletoe point; (B) Rounded-base point; (C) Thumbnail endscraper; (D) Montbani-retouched bladelet; (E) Trapeze.

Melsele, however, the sand dune was completely covered by peat after 4000 B.P. (at the latest); sherds are numerous. The ceramics are poorly fired and almost never decorated (Figure 5). They seemed to have been tempered with almost anything: bone, hematite, grog, grit, and flint microdebitage. Vessel forms are difficult to reconstruct but they include flat, round, and, most importantly, pointed-based forms (Van Roeyen and Van Berg 1989). These ceramics do not resemble in any way those of the LBK, except the exotic Limburg Ceramic which is also bone-tempered. However, the Melsele ceramics differ from those of LC in form, firing, and decoration. Nor do they resemble those of the later Rössian and Michelsberg Neolithic cultures. They are most comparable in form to those found at several of the Swifterbant sites in the Netherlands but are fired and tempered in a different way (de Roever 1979). In summary, the Melsele ceramics are unique but bear a family resemblance to those of the Swifterbant Ceramic Mesolithic.

As with the lithic-ceramic associations at Swifterbant and Weelde, the bioturbated nature of the archaeological layer leaves doubts concerning the contemporaniety of the ceramics and lithics. Fortunately, a bark-lined pit was found in 1990 which clearly pre-dates the bioturbation and contains distinct ashy layers with both Mesolithic flint artifacts and ceramic sherds, confirming their association. Bark samples are being dated by the Oxford Accelerator Unit.

There are several ^{14}C dates from Melsele but they give a confusing picture (one date was even in the future; apparently a mistake in the lab!). A recent date on some charcoal, or plant material carbonized by reduction, from the interface between the dune and the overlying peat was 5500 B.P., which is in accord with the few pieces of Middle Neolithic material present. However, this represents a terminal date; the date of the Mesolithic artifacts and ceramics is likely at least 500 years earlier. The lithics and ceramics are rather homogeneously dispersed through the 10 to 15 cm of the bioturbation horizon, implying they were present before the bioturbation began. R. Langohr (personal communication) estimated that the bioturbation would have required some 400–500 years to reach its final state of development, implying a date of at least 6000 B.P. for the bulk of the Melsele material. The Ceramic Mesolithic at Swifterbant is dated to 6300 B.P. (de Roever 1979; Price 1983; Whallon and Price 1976); the mean of the two dates from Weelde is 6350 B.P. A reasonable conclusion from this evidence is that some Late Mesolithic groups in Holland and northern

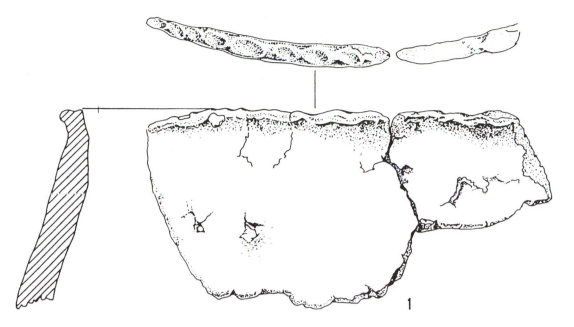

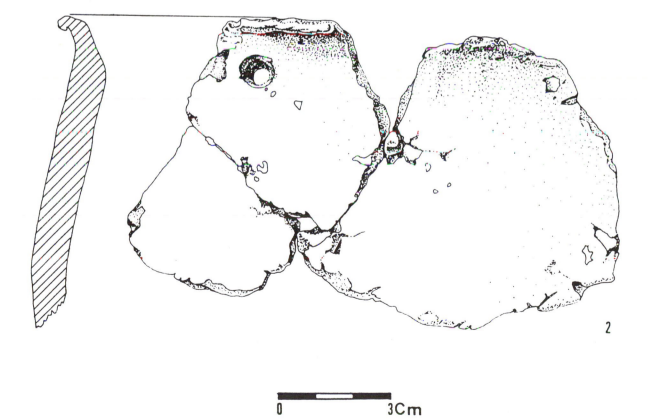

0   3Cm

Figure 5. Melsele Ceramics.

Belgium were making ceramics before 6000 B.P. Curiously, these ceramics bear almost no resemblance to those of the LBK, even though Melsele is only 42 km and Weelde is only 34 km from the nearest LBK settlements.

Although these ceramics are similar to those at Swifterbant, the lithic industries are not and instead are of a long-established local tradition. Thus, it seems that these Ceramic RMS groups in Belgium adopted, but did not copy exactly, the ceramic technology of the Late Mesolithic groups to the north while ignoring the fully Neolithic techniques of their LBK neighbors immediately to the southeast.

Late Mesolithic Subsistence

Because of the poor organic preservation at most Late Mesolithic sites, little can be said about subsistence patterns. There is scattered evidence that the Late Mesolithic groups hunted red deer, wild boar, aurochs, roe deer, beaver, and various other small game and birds. They are presumed to have fished but there is no direct evidence from our region. The fact that projectile points and armatures dominate the tool assemblages is indirect evidence that hunting and fishing were a major focus of Late Mesolithic life.

We are absolutely certain they gathered hazelnuts, since hazelnut shells are a ubiquitous feature at Mesolithic sites, very often the only organic remains recovered. It is presumed that they also consumed acorns, berries, seeds, and roots but, other than a few vetch seeds from Station Leduc (Gob 1990), these uses are unverified. The implements necessary for processing seeds and acorns, such as mortars, pestles, and grinding stones, are extremely rare. The impression is that plant foods, other than hazelnut, were not very important in Late Mesolithic economies. No evidence of domesticated plants has yet been found at any RMS site.

There is evidence at two Ceramic RMS sites, Weelde and Melsele, that domesticated cattle were being herded. Just under the peat at Melsele, bones do survive and Gauthier (personal communication) in a preliminary examination found, along with the remains of the usual wild prey, the bones of domesticated cattle and pig. Interestingly, the domestic cattle bones do not appear to be the small LBK breed but some larger strain. These impressions must be confirmed by the full faunal analysis now being conducted. Again, we find a "Neolithic" trait adopted by some RMS groups but derived from a source other than LBK. Pollen analysis at Weelde revealed elevated percentages of ivy which Vermeersch (1984:184) suggests might be the result of its use as winter fodder for domestic animals. In view of the many similarities

between Weelde and Melsele, this hypothesis is not improbable.

Late Mesolithic Settlement Pattern and Social Units

The general RMS settlement pattern, like that of the Late Mesolithic elsewhere, is the inverse of that of the LBK; concentrations of sites occur in major river valleys, around lakes and marshes, and on the Coversands to the north. Few sites are known from the loess proper, although this may be due to preservation and visibility factors (e.g., Gob 1984b:210; Bogucki 1988: 37). The finds of RMS artifacts incorporated in the fill of LBK pits at Place St. Lambert (Gob 1984a) and Oleye (see above) indicate that the settlement zones of these two cultures were not entirely mutually exclusive.

In Upper Belgium, Gob (1984b) notes that Mesolithic sites were usually placed on slopes overlooking valleys or other locations commanding good views. He notes, though, that a few late Mesolithic sites were located lower in the valleys. This might indicate a greater emphasis on aquatic resources. In the Coversands of Lower Belgium, sites were located on the tops or upper slopes of low dunes, i.e., at the highest places in a region of low topography.

RMS sites are small, light scatters of finds; features are rare, although the small scale of most excavations makes this a weak generalization. Also, it is quite possible that larger, more substantial sites lie buried by post-Neolithic alluvium in the river valleys. The general impression, however, is that the settlements systems were rather mobile, that storage of food was unimportant, and that population densities were low, particularly in contrast to the LBK. There are, however, some persuasive arguments that during the Late Mesolithic in northwestern Europe, in contrast to the earlier Mesolithic, population density was higher, mobility was lower (Barker 1985: 165-167), exchange systems more extensive, and social boundaries were more sharply defined (Gendel 1984)—all indications of increasing socioeconomic complexity.

Late Mesolithic—Discussion

The Final RMS Mesolithic, in almost every aspect, is a direct continuation of the earlier RMS Mesolithic. In lithic technology, there is a high degree of stylistic continuity in technique, form, and even raw material preferences from the earlier Mesolithic. The economy was based, like that of the earlier Mesolithic, on the hunting of forest species, especially red deer and wild pigs, fishing, and gathering of hazelnuts. Settlements were still quite mobile and oriented towards the more open environments of major river valleys and wet lowlands. Some weak evidence implies that during the Late Mesolithic, population density, sedentism, and social complexity were increasing. There is growing evidence that some RMS groups adopted ceramics and domesticated animals before 6000 B.P. but the precise date is still uncertain.

Farmer-Forager Interactions in Northwest Europe

There is no area of archaeological theory that is sunnier in climate or more innocent of cynicism than that concerned with farmer-forager interaction (Moore 1985; Gregg 1988; Bogucki 1988; Dennell 1985). The key concept in all of this literature is "mutual benefit." These theorists have argued that it was of benefit to farmers to have foragers around and to the foragers to have farmers nearby. These benefits accrue from the exchange of agricultural for "forest" products, foragers' labor, or "information," perhaps supported or enhanced by the exchange of mates. Some of these arguments even imply that forest foods or labor of foragers were, at times at least, a necessity for the farmers (Bogucki 1988; Gregg 1988). So useful and agreeable do these interactions appear, at least in theory, one would expect them to have occurred regularly and to have been sustained over long periods.

If such interactions occurred between LBK and RMS, there are several consequences we might expect: (1) common finds of wild plants and animals at LBK villages, and of domesticated plants, especially grain, and animals at RMS sites; (2) finds of items of LBK material culture at RMS sites, and vice versa; (3) with "information" and marriage exchanges, mutual acculturation should be expressed in the adoption of or blending of stylistic traits; (4) a shifting of RMS settlement location towards LBK settlements (Gregg 1988: 234), or (5) an increase in sedentism and social complexity for RMS settlements (Moore 1985: 107). In short, RMS and LBK sites should gravitate towards one another and their archaeological characteristics should become less distinct.

Evidence for LBK-RMS exchange of food is weak to nonexistent. The remains of wild animals at LBK sites are extremely rare and there is no reason to suppose that such small numbers were not obtained directly by the LBK folks. The only bones of domestic cattle found at a RMS site are apparently of a non-LBK breed. There is no evidence of any agricultural plant foods at any RMS site. Hazelnuts are common at LBK sites but pollen studies indicate their abundance in the immediate vicinity of the villages. If the LBK and RMS were trading food, which is doubtful, it was a very minor activity.

Given the eager acceptance by recent hunter-gatherers of metal and ceramic containers, it is surprising that no finds of LBK ceramics have ever been made at RMS sites. The few ceramics associated with RMS lithics are clearly of non-LBK types. Unless we are willing to accept Modderman's hypothesis, despite all the evidence to the contrary, that LC is a Mesolithic production, there is no evidence of Mesolithic ceramics at LBK sites. There are a number of instances of Danubian points found in RMS assemblages (De Laet 1982: 208). However, since triangular, thinned-based points, very similar to Danubian points, are found throughout the Mesolithic, some of these identifications may be suspect. Finds of the very distinctive LBK adzes at RMS sites do occur but are much more rare than finds of supposed LBK points. Finds of RMS implements at LBK sites are rare and primarily involve mistletoe points (De Laet 1982:207). The more varied RMS finds at Oleye and Place St. Lambert, as already mentioned, are indicative of replacement, not contact. To sum up, if LBK and RMS were exchanging anything, it was projectile points and adzes, i.e., weapons; there is no reason to assume that such "exchanges" were non-violent.

There is no evidence of change in LBK lithic technology during its span in our area; it certainly did not become more Mesolithic. The major western LBK innovation in ceramics is the increasing use of comb decoration but it is difficult to imagine that this was learned from the essentially *aceramic* RMS. Some RMS groups in northern Belgium adopted ceramics and livestock but these are not based on LBK models. RMS lithics, other than the rare supposedly Danubian points, show no evidence of LBK influence. If they traded mates, the participants must have been singularly deficient in long-term memory. If they exchanged information, it was not about anything important.

There are arguments that RMS was becoming more socially complex, but these refer to the whole span of RMS not just the period of potential LBK contact. Some the latest RMS sites (ca. 6300 B.P.) seem just as ephemeral as those preceding the LBK colonization (e.g., Maarheeze, Brecht-Mordenaarsven 2, etc.). As for LBK villages attracting RMS settlement, the opposite seem to be the case. Where no major geographic boundary exists, such as the trench of the Sambre-Meuse Rivers, an unsettled no-man's-land, some 20-30 km wide, seems to have divided LBK from RMS settlement zones (Figure 6). It is much easier to argue from these data that RMS and LBK repelled, rather than attracted, one another. Not only is there little support for the hypothesis of peaceful interactions between RMS and LBK, the simplest reading of the evidence is that they had little to do with one another.

There is even evidence that relations were not just chilly but actively hostile. In the Upper Geer region, six LBK sites have recently been excavated. Four of these (Darion, Oleye, Waremme-Longchamps, Vaux-et-Borset) were at the limits of the LBK settlement zone facing the no-man's-land beyond; all four were extensively fortified. The two sites inside the zone gave no evidence of fortification (Hollogne-sur-Geer, Vieux Waleffe). The threat, then, came from beyond the no-man's-land; the only people out there, as far as we know, were the RMS groups. In fact, if we look at the distribution of all LBK "enclosures" across western Europe (Figure 7), they cluster at the edges of the LBK settlement zone. Again, the implication is that threat was external; the only viable candidates for this threat are hostile Mesolithic foragers.

Conclusions

An agricultural way of life came to western extension of the North European Plain by two very different routes. One involved the migration and colonization of the forested loess by foreign LBK farmers who cleared the forest for their fields and pastures, substituted their alien domestic plants and animals for the native flora and fauna, and brought with them a totally alien way of life. Within the zone of their settlement, they apparently completely and permanently replaced the indigenous RMS Mesolithic culture. As for the RMS groups living beyond the area of LBK settlement, these they avoided or even fought. While the RMS seems to have gained or learned little from the LBK, some groups of the former did adopt elements of a neolithic way of life, specifically ceramics and stock-rearing.

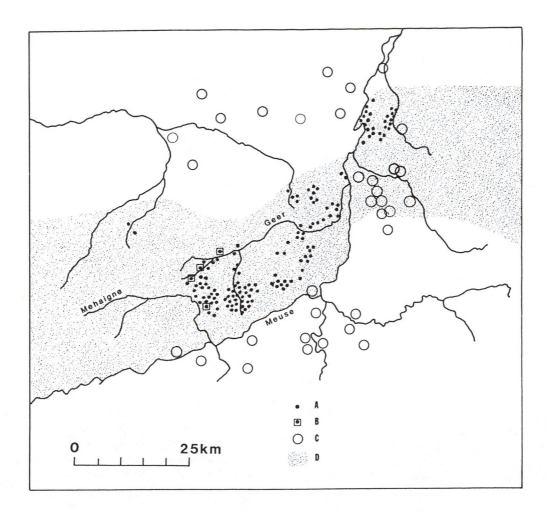

Figure 6. LBK and Final Mesolithic sites in Eastern Belgium. (A) LBK site; (B) Fortified LBK site; (C) Final Mesolithic site; (D) Loess.

These novelties were not acquired from the neighboring LBK but from similarly acculturating Mesolithic groups such as those at Swifterbant to northeast.

Acculturation does not necessarily follow the most direct or obvious routes. The adoption of "horse culture" by 18th and 19th century Native Americans may parallel the LBK-RMS case. Most North American tribes did not adopt the complex of traits associated with the use of horses directly from the Europeans with whom they were in immediate contact, but rather from other Indian tribes whose way of life was more similar to their own and whose breeds of horses could be more easily incorporated into their existing economy. The Plains tribes did not adopt the wagon, potato, wheat, or plow from the alien Euro-Americans. But with the horse, they created a new way of life, one that still excites our admiration and romantic imaginations in a way that the life of the stolid farmers who displaced them cannot. It is sometimes difficult to remember that it was the better organized, unheroic, yet relentless farmers who eventually remained.

Acknowledgements

The results reported here are the product of a five-year collaboration with Daniel Cahen, Director of the Institut Royal des Sciences Naturelles de Belgique. This work was supported by grants from the FNRS of Belgium and the NSF of the USA (BNS-8616463 and 8820404). I also owe much to the work of my colleagues: Drs. Langhor and Gauthier of Gent University, Drs. Heim and Gilot of the University of Louvain, Mssrs. Van Hove and Van Roeyen of the Archeologische Dienst Waasland, and Dr. Van Berg of the IRSNB.

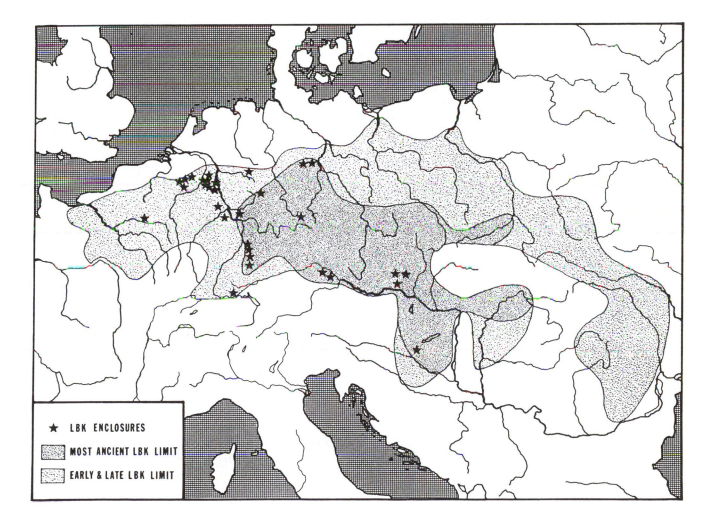

★ LBK ENCLOSURES

▨ MOST ANCIENT LBK LIMIT

▦ EARLY & LATE LBK LIMIT

Figure 7. LBK "enclosures" and limits of LBK settlement zone (After Lüning 1988).

References Cited

Anthony, D.
  1990 Migration in Archaeology: the Baby and the Bathwater. *American Anthropologist* 92:895-914.
Bakels, C.
  1982 The Settlement System of the Dutch LinearBandKeramik. *Analecta Praehistorica Leidensia* 15:31-44.
Barker, G.
  1985 *Prehistoric Farming in Europe*. Cambridge University Press, New York.
Bogucki, P.
  1988 *Forest Farmers and Stockherders*. Cambridge University Press, New York.
Cahen, D. and E. Gilot
  1983 Chronologie Radiocarbone du Néolithique Danubien. In *Progrés Récent dans l'êtude du Néolithique Ancient*, edited by S. De Laet, pp. 21-40. De Tempel, Brugge.
Constantin, C.
  1985 *Fin du Rubané, céramique du Limbourg et Post-Rubané*. BAR International Series 273. BAR, Oxford.
De Laet, S.
  1982 *La Belgique d'avant les Romains*. Universa, Wetteren.
Dennell, R.
  1985 The Hunter-Gatherer/Agricultural Frontier in Prehistoric Temperate Europe. In *The Archaeology of Frontiers and Boundaries*, edited by S. Green and S. Perlman, pp. 113-139. Academic Press, New York
Desse, J.
  1984 Les Restes des poissons dans les fosse omaliennes. In *Les Fouilles de la Place St-Lambert á Liäge*, edited by M. Otte, pp. 239-240. Études et Recherches Archéologiques de l'Université Liége, No. 18.
de Roever, J.
  1979 The Pottery from Swifterbant—Dutch Ertebølle? *Helinium* 19:13-27.
Geddes, D.
  1985 Mesolithic Domesticated Sheep in Western Mediterranean Europe. *Journal of Archaeological Science* 12:25-48.
Gendel, P.
  1982 The Distribution and Utilization of Wommersom Quartzite during the Mesolithic. In *Le Mésolithique entre Rhin et Meuse*, edited by A. Gob and F. Spier, pp. 21-33. Société Préhistorique Luxembourgeoise, Luxemburg.
  1984 *Mesolithic Social Territories in Northwestern Europe*. BAR International Series 218. BAR, Oxford.
Gilot, E.
  1984 Datations radiométrique. In *Peuples Chasseurs de la Belgique Préhistorique dans leur Cadre Naturel*, edited by D. Cahen and P. Haessaerts, pp. 115-125. Institut Royal des Sciences Naturelles de Belgique, Brussels.
Gob, A.
  1984a L'industrie mésolithique. In *Les Fouilles de la Place St-Lambert -Liege 1*, edited by M. Otte, pp. 147-152. Etudes et Recherches Archéologique de l'Université de Liége 18.
  1984b Les industries microlithique dans la partie sud de la Belgique. In *Peuples Chasseurs de la Belgique Préhistorique dans leur Cadre Naturel*, edited by D. Cahen and P. Haessaerts, pp. 195-210. Institut Royal des Sciences Naturelles de Belgique, Brussels.
  1990 Du Mésolithique au Néolithique en Europe nord-occidentale: un point de vue de mésolithicien. In *Rubane et Cardial*, edited by D. Cahen and M. Otte, pp.155-160. Études et Recherches Archéologiques de l'Université Liége 39.
Gob, A. and F. Jacques
  1985 A Late Mesolithic Dwelling Structure at Remouchamps, Belgium. *Journal of Field Archaeology* 12:163-175.
Gregg, S.
  1988 *Foragers and Farmers*. University of Chicago Press, Chicago.

Hiem, J.
  1983 Apport Récents de la Paléobotanique — la Connaissance de l'Importance des Activities Culturales (Agricoles) des Néolithiques Anciens entre Rhin et Seine. In *Progrés Récent dans l'étude du Néolithique Ancient*, edited by S. De Laet, pp. 62-70. De Tempel, Brugge.
Keeley, L., and D. Cahen
  1989 Early Neolithic Forts and Villages in NE Belgium: a Preliminary Report. *Journal of Field Archaeology* 16:157-176.
  1990 Village Specialization in the Early Neolithic of NW Europe. Paper read at Annual Meeting of Society for American Archaeology, Las Vegas. NV.
Lüning, J.
  1982 Research into the Bandkeramik Settlement of the Aldenhovener Platte in the Rhineland. *Analecta Praehistorica Leidensia* 15:1-29.
  1988 Zur Verbreitung und Datierung Bandkeramischer Erdwerke. *Archaeologisches Korrespondenzblatt* 18:155-158
Milisauskas, S.
  1978 *European Prehistory*. Academic Press, New York
Milisauskas, S., and J. Kruk
  1989 Neolithic Economy in Central Europe. *Journal of World Prehistory* 3:403-446.
Modderman, P.
  1981 Céramique du Limburg: Rhénanie, Westphalie, Pays-Bas, Hesbaye. *Helinium* 21:140-160.
Moore, J.
  1985 Forager/Farmer Interactions: Information, Social Organization, and the Frontier. In *The Archaeology of Frontiers and Boundaries*, edited by S. Green and S. Perlman, pp. 93-112. Academic Press, New York
Otte, M., and L. Keeley
  1990 The Impact of Regionalism on Palaeolithic Studies. *Current Anthropology* 31:577-582
Price, T. Douglas
  1983 Swifterbant, Oost Flevoland, Netherlands: Excavations at the River Dune sites, S21-S24, 1976. *Palaeohistoria* 23:75-104.
Sherratt, A.
  1976 Resources, Technology and Trade: an Essay in Early European Metallurgy. In *Problems in Economic and Social Archaeology*, edited by G. Sieveking, I. Longworth, and K. Wilson, pp. 557-582. Duckworth, London.
Van Berg, P.-L., and D. Cahen
  1992 Relations sud-nord au Néolithique Ancien en Europe occidentale. In *Actes du Colloque interrégional sue le Néolithique*. CNRS, Metz, in press.
Van Roeyen, J.-P. and P.-L. Van Berg
  1989 Les chasseurs ceramisés du Pays de Waas. *Notae Praehistoricae* 9:31-32
Vermeersch, P.
  1984 Du Paléolithique final au Mésolithique dans le nord de la Belgique. In *Peuples Chasseurs de la Belgique Préhistorique dans leur Cadre Naturel*, D. Cahen and P. Haesaerts, editors, pp. 181-193. Institut Royal des Sciences Naturelles de Belgique, Brussels.
Wahl, J., and H. G. König
  1987 Anthropologisch-Traumologische untersuchung der Menschlichen Skelettreste aus dem Bandkeramischen Massengrab bei Talheim, Kries Heilbronn. *Fundberichte aus Baden-Wurttemberg* 12:65-193.
Whallon, R. and T. D. Price
  1976 Excavations at the River Dune Sites S11-13. *Helinium* 16:222-229.
Whittle, A.
  1987 Neolithic Settlement Patterns in Temperate Europe: Progress and Problems. *Journal of World Prehistory* 1:5-52.

8

The Final Frontier:
Foragers to Farmers in
Southern Scandinavia

T. Douglas Price
Anne Birgitte Gebauer
University of Wisconsin–Madison

Agriculture spread across prehistoric Europe between 6,000 and 3,000 b.c. in the form of plants and animals originally domesticated in the Near East. Until the last ten years or so, this introduction was thought to reflect the spread of foreign colonists bearing ceramic containers and domesticated plants and animals and bringing permanent villages, new architecture, storage facilities, long-distance trade, and elaborate burial rituals. The indigenous hunter-gatherers of the continent were thought to have been largely aceramic, residentially mobile, dependent on wild foods, socially amorphous, and eventually overwhelmed. Today, however, the "neolithic revolution" in much of Europe is regarded as an inside job, resulting from indigenous local groups "borrowing" Neolithic materials and practices (e.g., Price 1987, Runnels and van Andel 1988, Whittle 1987, 1988). With only a few exceptions (e.g., Keeley, this volume), the first farmers of Europe were the last hunters.

The question of why these foragers adopted farming remains unresolved. Information from northern Europe, because of both its abundance and its quality, is particularly well suited to addressing this question. The establishment of a food-producing economy in southern Scandinavia appears to have been a slow and gradual process. During the late Mesolithic certain ideas or actual objects such as pottery, stone and antler axes, and bone combs were obtained in trade from farmers to the south after 3600 b.c. [uncalibrated radiocarbon years B.C.]. Farming populations at that time were present only two hundred kilometers or so away in Poland and northern Germany (Figure 1). Despite such substantial evidence of contact with these Neolithic communities, agricultural foodstuffs were among the last "Neolithic" items to be brought into Scandinavia. Evidence for farming appeared suddenly across Denmark and southern Sweden around 3100 b.c. with the arrival of domesticated plants and animals, a range of new ceramic containers, large burial monuments, and new prestige items. Another 500 years went by before a major agricultural expansion around 2600 b.c. made farming the established way of life in southern Scandinavia. The imperviousness of the late Mesolithic people to Neolithic influence in their economy delayed the adoption of agriculture by almost a 1000 years. The only reasonable explanation for this delay is the presence of successful hunter-gatherers who had little use for domesticated plants and animals.

Thus, the focus for this discussion is the question of why these foragers in southern Scandinavia became farmers. The specific issues to be pursued concern the nature and the causes of the transition from hunting and gathering to agriculture. The paper is divided into four parts. Part I will provide an introduction to the traditional chronology and characteristics of Mesolithic and Neolithic southern Scandinavia between 4500 b.c. and 2200 b.c. Part II is a description of four stages in the transition which we have designated as Mesolithic, Last Hunters, First Farmers, and

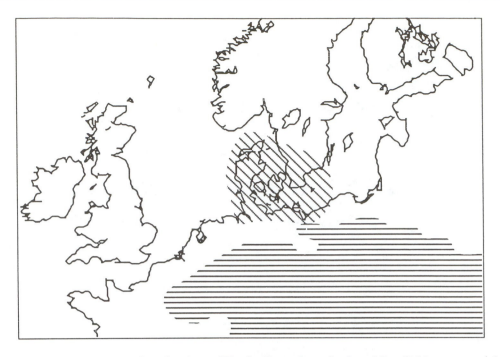

Figure 1. The approximate distribution of Ertebølle and equivalent Mesolithic groups (above) and Linearbandkeramik Neolithic groups (below) in northwestern Europe around 4000 b.c.

Neolithic. Part III is a presentation of the current theoretical debates regarding the Neolithic transition to agriculture in southern Scandinavia. Part IV is a discussion of some of the important variables in this transition to agriculture, including climatic and environmental change, resource availability and food stress, population size and density, circumscription and territorial behavior, and social differentiation. It is our intent to evaluate these variables in terms of the role they may have played in the spread of agriculture into this area.

Southern Scandinavian Prehistory, 4500–2200 b.c.

The chronology of the transition to agriculture is well defined in southern Scandinavia and there is reasonably good evidence from a large number of sites. The period of concern extends from 4500 b.c. to 2200 b.c., encompassing the late Mesolithic (*Ertebølle - EB*) and the early Neolithic (*Tragtbæger* [*TRB*] or *Funnel Beaker*) in this area (Figure 2).

Ertebølle

The Ertebølle, 4500–3100 b.c., is found throughout southern Scandinavia and across the coastal areas of northern Germany and Poland. Three distinct phases of the Ertebølle are recognized in southern Scandi-

navia (Vang Petersen 1984). The early Ertebølle (4500–4000 b.c.) is characterized by small oblique points, an absence of stone scrapers and soft hammer flaking, and core axes. The middle Ertebølle (400–3500 b.c.) is marked by the appearance of symmetrical, transverse points and concave end scrapers on blades, the re-emergence of soft hammer flaking, and flake axes replace core axes. Crude, thick-walled pottery appears in this phase of the Ertebølle. The projectile points of the later Ertebølle (3500–3100 b.c.) are narrow and transverse, both core and flake axes are present, and small, ceramic lamps appear at this time. Significant regional variation in artifact types and styles is documented during the later half of the Ertebølle; differences between eastern and western Denmark and southern Sweden, and among smaller areas within Zealand have been reported (Andersen 1980; Larsson 1990; Vang Petersen 1984).

The Ertebølle represents the culmination of several trends in the Mesolithic. Technological elaboration accompanied the development of the Mesolithic in southern Scandinavia (Blankholm 1987; Price 1985, 1986). More artifact types and facilities, and more complex facilities, are known from later than from earlier periods; previous forms became more functionally specific. A wide array of wood, bone, and antler tools was in use by the Ertebølle period. Watercraft in the form of dugout canoes up to 10 m in length provided for the movement of people and goods.

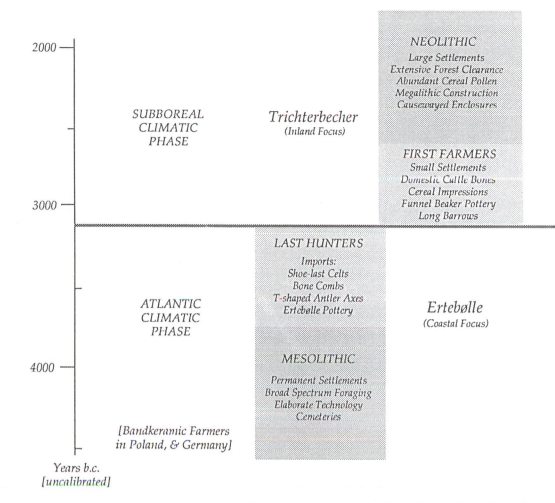

Figure 2. The chronology and terminology of the transition to agriculture in southern Scandinavia.

Technological changes of such magnitude are correlated with resource utilization. An intensification in food procurement can be traced through the Danish Mesolithic. The presence of shell middens, the diversity of extraction camps, the faunal remains of a wide range of marine fish and mammals, including seals, dolphins, and whales, and the utilization of "species-specific" trapping stations, all combine to demonstrate the diversity of the subsistence base. The number of species represented at Ertebølle sites is some 50% greater than in the earlier Maglemosian.

A number of different types of sites are known from the Ertebølle including (1) coastal occupations, containing both marine and terrestrial fauna, with or without associated shell middens. The shell middens appear to be largely long-term, episodic accumulations of seasonal activities (Andersen and Johansen 1987). (2) Smaller, seasonal coastal sites with a more specific procurement focus—deep water fishing, sealing, or fowling for migratory species such as swans. (3) Inland trapping stations with large numbers of intact carcasses from fur-bearing animals such as pine marten. And perhaps (4) inland lakeside settlements that may be year-round occupations.

Recent investigations of Mesolithic diet in southern Scandinavia (Price 1989, Tauber 1981) have altered a bias toward terrestrial foods that dominated earlier views. The importance of marine resources in the Mesolithic in northern Europe has been seriously underestimated. Tauber (1981) used carbon isotopes in human bone to examine the importance of marine resources in the diet of the inhabitants of Mesolithic and early Neolithic Denmark. Mesolithic hunter-gatherers in southern Scandinavia were assumed to have been largely dependent on terrestrial resources. Although a variety of marine foods, including fish, seal, and whale bones, were found on sites from this period, their contribution to the diet was thought to be relatively small. However, analysis of carbon isotope ratios in the Mesolithic hunter-gatherers were found to be comparable with those of Greenland Eskimo where marine foods contribute more than 75% of the diet.

Thus, it appears that Mesolithic hunter-gatherers only supplemented their diet with terrestrial foods. Carbon isotope ratios in the Neolithic farmers were clearly representative of terrestrial foods, documenting both a dramatic shift away from the sea and the importance of domesticated plants and animals in the diet. Archaeological evidence such as substantial quantities of fish bone at coastal sites and the presence of marine mammal bones at inland sites in Jutland (Andersen 1975) reinforces this impression of the importance of marine resources in the diet of the Mesolithic inhabitants of northern Europe.

Year-round occupation at both coastal and inland sites appears likely during the Ertebølle and this practice has antecedents in the preceding Kongemose period (Larsson 1988, 1990; Rowley-Conwy 1983). Recent carbon isotope analysis of dog bones from inland Ertebølle sites reveals no marine foods in the diets of domestic dogs, suggesting that these inland groups did not move to coastal areas during some seasons of the year (Noe-Nygaard 1988), again emphasizing the likelihood of sedentary residence. Noe-Nygaard (1983) has also documented the importance of freshwater aquatic resources at these inland sites.

The occurrence of cemeteries in Zealand and southern Scania at this time complements this picture of more sedentary residence and suggests increased social and ritual complexity. Cemeteries are a recently discovered hallmark of the Mesolithic, unknown in northern Europe prior to 1975. A few isolated graves had been reported in the literature but it was not until the discoveries at Vedbæk, Denmark (Albrethsen and Brinch Petersen 1976), and at Skateholm in southern Sweden (Larsson 1988), that the presence of substantial graveyards was recognized (Figure 3). At Vedbæk in northeastern Zealand, a Mesolithic cemetery was uncovered in 1975 during the construction of a school. The cemetery is dated to approximately 4000 b.c. and contains the graves of at least 22 individuals of both sexes and various ages. Powdered red ochre was used to adorn individuals in many of the graves. Racks of red deer antler were placed with elderly individuals; males were buried with flint knives; females often were interred with jewelry of shell and animal teeth. The cemetery also contained rather dramatic evidence for conflict and warfare. The simultaneous burial of three individuals in a single grave—an adult male with a lethal bone point through his throat, an adult female, and a child—suggests the violent death of all three.

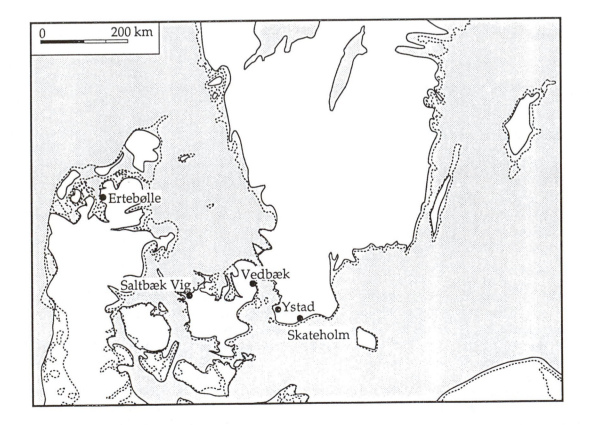

Figure 3. The location of sites and areas in southern Scandinavia mentioned in the text.

Excavations at Skateholm in southern Sweden have uncovered at least three cemeteries within the same former estuary, dating from the late Mesolithic (Larsson 1988). Skateholm I contained at least 57 graves with some 62 individuals, as well as 8 dog graves. Skateholm II, only partially excavated, held at least 22 graves; an unknown number of graves were found nearby at Skateholm III some years ago. Almost every imaginable type of burial has been found here, including cremations and inhumations. The inhumation graves show individuals in a variety of positions, supine, sitting, extended, flexed, and more.

Grave goods and food accompany a number of the burials. Both position in the grave and the type of grave goods appear to vary with the age and sex of the individual, but there is no evidence of status differentiation. However, social positions based upon attained status within Ertebølle society may be indicated by extended exchange networks and the acquisition of prestige items. Imports from farming communities to the south include the knowledge of pottery production and commodities such as T-shaped antler axes, bone combs, and rings. The presence in Ertebølle context of shaft-hole axes—the so-called *Schuhleistenkeile*—made of amphibolite from Southeastern Europe provide further evidence of this contact (Fischer 1982). In a Neolithic context, these axes appear to be status symbols carried by certain males (Sherratt 1976). The use of these axes as weapons is documented in the 35 individuals in a mass grave at Talheim, Germany, killed by blows to the head with such an axe (Wahl and König 1987; see also Keeley, this volume). Regardless of their use among Ertebølle people, the import of prestige items like *Schuhleistenkeile* suggest that Ertebølle communities had a social structure where prestige was important. Participation in exchange networks with Neolithic groups may also indicate a social structure not greatly different from those of the farming communities. Finally, in addition to commodities, these exchange links functioned to bring a variety of information and skills to southern Scandinavia (Madsen 1986).

Tragtbæger

The earliest radiocarbon-dated evidence for the Neolithic comes from long barrows with timber burial chambers, bog site offerings, and a flint mine. The number of initial Neolithic sites is limited, impeding an understanding of the transition process itself. The earliest radiocarbon dates for actual settlements are 100 years younger, often at sites in continuous use from the Ertebølle period. Evidence from several sites documents a continuity between Ertebølle and TRB,

but there is no evidence for contemporaniety between the two (Nielsen 1985).

The settlement pattern of the earliest TRB is one of residential sites combined with seasonal hunting locations (Madsen 1982, Madsen and Jensen 1982). Residential sites are small (500—600 m²) compared to the Ertebølle period, with a thin cultural layer suggesting that co-resident groups in the early Neolithic were smaller and perhaps more mobile. Sites are located further from the coast than during the Ertebølle, often near lakes or streams where fresh water was easily obtainable and conditions for grazing were favorable.

Several hunting sites are known between 3100–2600 b.c. containing either a mixture of wild and domestic fauna or purely wild species (Skaarup 1973, Madsen 1986). These sites continued in use from the Ertebølle period. Activities at these seasonal camps included hunting for red deer, wild boar, baby seals, ducks, swans, collecting shellfish, berries and nuts, and fishing, as well as the summer grazing of domesticated cattle and sheep. The number of hunting sites decreases dramatically around 2600 b.c.

Substantial residential sites with thick cultural layers began to appear on the coast as well as inland around 2600 b.c. In some areas the distribution of permanently inhabited sites at 2 km intervals can be observed, each with one or more megalithic tombs and offering sites (Skaarup 1985). Houses from this period measure 8–16 m in length and are ca. 6 m wide (Larsson and Larsson 1986) and settlement size ranges up to 40,000 m² (Skaarup 1985). Residential settlements likely represent individual homesteads or small hamlets.

The trend toward denser and larger habitation units continues through the late Funnelbeaker period, 2450-2200 b.c. Rich residential sites dependent on cereal and cattle vary in size from 70,000 to 300,000 m². In some areas these sites are found as close as 1 km apart (Skaarup 1985). A few houses, 13–18 m long and 6–7 m wide, are known from this period (Nielsen 1985). Hunting and gathering were now incorporated with activities at the residential sites, but special camps for fishing and sealing, along with flint knapping sites for axe production, are also known.

The Tragtbæger settlement pattern has a more inland focus than the Ertebølle habitation. As noted above, carbon isotope analysis of Neolithic skeletons indicates a sharp decrease in the consumption of marine foods in favor of a greater dependence on terrestrial resources (Tauber 1981). It is clear, however, that hunting, fishing and gathering still played an important role in the diet during the first 500 to 600 years of the Neolithic and continued to provide a supplement to the economy throughout the TRB. However, Ertebølle sites and other hunting stations probably were only used for the seasonal procurement of special resources rather than as residential sites providing the major portion of the diet like in the Ertebølle.

Cereal cultivation and animal husbandry were likely introduced simultaneously. The domesticates introduced to southern Scandinavia from continental Europe included emmer wheat, einkorn wheat, club wheat, and barley along with cattle, pigs, and sheep/goat (Aaris-Sørensen 1988, Helbæk 1955, Hjelmquist 1979, Nielsen 1985). Cattle, pig, sheep/goat, and dog are present at most early Neolithic sites, but the faunal remains are too few to determine the relative importance of the various species during the initial period of farming. Later evidence indicates that cattle were the most important livestock and became more common through the Funnelbeaker period, comprising more than 80% of the domestic animals at some late sites (Andersen 1981, Madsen 1982). The primary importance of cattle was for meat; milk production and the use of animals for transportation or draught seems less important. Pigs were the second most common domestic species, clearly more important at inland sites where they may have roamed the forest. Pigs and sheep also were slaughtered for meat. Despite the presence of sheep, it is unclear whether wool and textiles were actually produced in the TRB period. Numerous cord impressions are seen on the pottery, but only a couple of spindle whorls are known. Fur and leather may have been preferred for garments as in the Mesolithic.

Around 2600 b.c. extensive forest clearance, together with evidence for more cereal growing and pasture land, indicate an agricultural expansion (Iversen 1941). Wheats comprise 96% of the cereals at that time, but barley increases to 22% of the cereal remains at the later Funnelbeaker sites. This shift may be due to a natural selection since barley is better adapted to the Scandinavian climate. Cereal samples appear to be monocultural and contain only few weeds. It is not clear whether this is due to seeding or harvest methods, perhaps combined with sorting and screening of the seed (Andersen 1981, Jørgensen 1977).

The importance of animal husbandry versus cereal growing during the TRB is unknown. Pollen analyses show only limited evidence of pasture and cereal cultivation during the first 500 years of the Neolithic. Based on the location of settlements, near mixed resources and freshwater, Madsen (1982) has suggested that the early farmers relied on pigs roaming the forests and cattle grazing the open areas along streams and lakes. A heavier reliance on cattle using leaf foddering in byres has been suggested by Rasmussen (1990). There is a clear trend toward a greater reliance on farming and an intensification of food production after 2800 b.c. when a number of new sites appear and megalith construction begins, i.e., some 300 years after the introduction of domesticates. Major agricultural expansion does not take place before 200 years later, around 2600 b.c.

The introduction of farming was accompanied by a number of technological innovations. Large polished flint axes were produced for forest clearing and timber work; a related flint mining industry supplied the raw materials. Grinding stones for processing cereals appear sometimes during the early Neolithic. A thin-walled type of pottery and new pot shapes suggest new methods of food preparation as well as the presence of different kinds of food and drinks (Nielsen 1987). New weapons and male status symbols were introduced such as groundstone battle axes, mace heads, copper axes, and flint daggers. Personal ornaments made of amber and copper appear. However, the manufacture of unpolished flint tools such as scrapers and arrowheads, as well as the basic pottery tradition, appears to be derived from the preceding Ertebølle technology (Nielsen 1983).

New social and religious patterns emerged in the TRB. An increasing amount of accumulated surplus was invested in ritual activities culminating in the period between 2600–2450 b.c. Enormous energy was invested in an ancestor cult. Early Neolithic burials were either in simple inhumation graves or in long barrows framed by timber palisades and containing wooden burial chambers (Figure 4) (Madsen 1979; Gebauer 1988a). Long barrows grew sequentially in size as compartments were added for new burials. Later, around 2800 b.c., the wooden burial constructions were replaced by megalithic chambers built of moraine boulders. In fact, most of thousands of megalithic tombs throughout southern Scandinavia were constructed between 2800–2450 b.c. The earliest megaliths, the so-called dolmens, consisted of a stone cist, covered by round or rectangular earthen mounds and circumscribed by a row of boulders. Later, around 2600 b.c., a number of larger chambers, accessed through doorways or passages, were added to the types of megalithic tombs. During the later TRB old tombs were reused and new ones not built (Ebbesen 1975, 1978, 1979, Skaarup 1985).

The megalithic tombs, like simple inhumation graves, were initially built for one funeral (Thorsen 1981). Part of the burial ritual appears to have been a continuation from the Ertebølle tradition. Both sexes and all age groups are represented in the burials with no apparent differentiation in the ritual (Madsen 1989). Several different types of graves or tombs were used at any given time, likely indicating some kind of social differentiation. The megalithic tombs were not open to

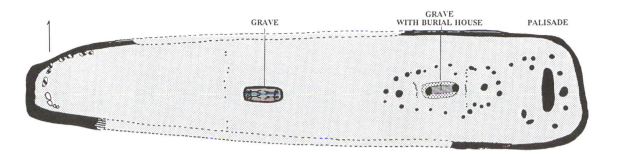

Figure 4. The earthen long barrow at Bygholm Nørremark included a grave placed in a wooden burial chamber. Early Neolithic pottery associated with this burial was found at a wooden palisade immediately to the east. The wooden chamber containing four persons was probably added later. A mound covered both graves and was surrounded by a wooden palisade. Such tombs document the elaborate burial of a few individuals in the Early Neolithic. (Courtesy of Torsten Madsen 1979)

everyone for burial. The function of the monuments as a family tomb symbolizing the rights and status of the lineage may have been as important as the commemoration of a high status individual.

Tombs and non-megalithic burials are often found in clusters around settlements. Burial monuments probably symbolized the territorial rights of the local inhabitants, using their ancestors as a means of legitimating rights to land and other resources. Clusters of tombs are found at 2 km intervals perhaps reflecting the diameter of local group territories (Skaarup 1985). Increased local style differences in the pottery reflect greater regional differentiation during the period of megalithic construction between 2800–2450 b.c. (Gebauer 1988b).

Votive offerings associated with the local community took place at nearby wetland areas between 3000–2500 b.c. In the simplest form these offerings included only a pot, presumably filled with food. More elaborate events include the construction of wooden platforms off the shoreline and sacrifices of humans as well as cattle and various artifacts (Figure 5). Repeated sacrifices appear to take place at the same bog and there is a trend toward larger bog offerings over time during the early Neolithic (Nielsen 1983). Distinctive evidence of communal feasting, where humans made up part of the menu, are reflected in the heaps of smashed bones found at some sites (Becker 1948, Bennike and Ebbesen 1987, Rech 1979, Ebbesen 1982, Skaarup 1985).

Causewayed enclosures may reflect this tradition of early Neolithic bog offerings, only in a larger and more elaborate form. Causewayed enclosures were constructed between 2600–2450 b.c. on promontories, surrounded by streams and wetlands on two or three sides (Andersen 1988, Madsen 1989). The remaining perimeter was enclosed by a palisade and a system of interrupted ditches. Activities were centered around these ditches. In the ditches are found whole vessels, as well as deliberately broken pottery, layers of animal bones, charcoal, and ash, skulls of dogs, and a few human remains such as a lower jaw—probably the remains of major offerings as well as feasting. Several hundred people must have been involved in the construction of these sites and the subsequent celebrations. These enclosures may have acted as integrative centers for large groups, combining ceremonial functions with social needs such as leveling conflicts, negotiating the exchange of goods, arranging marriage partners, and the like.

An intensive network of exchange promoted the trade of amber as well as flint and axes within southern Scandinavia. From around 2600 b.c. there is a movement of amber from western to eastern Denmark, suggesting an intensification of inter-regional trade relations. A number of organic products were probably included in the trade such as fur, feathers, lamp oil from seals, and honey. Far reaching connections to south-central Europe are reflected in the import of copper ornaments and jewelery, axe blades, and one battle axe, a model for the Scandinavian axes (Randsborg 1979). Numerous caches of flint axes, amber beads, and copper may represent safe deposits of valuable trade items. In some cases, however, the method of burial and the artifacts themselves suggest that items were not meant to be retrieved by humans (Nielsen 1977, Rech 1979).

Stages in the Transition to Agriculture

The transition to agriculture in southern Scandinavia was a very gradual process lasting at least 1000

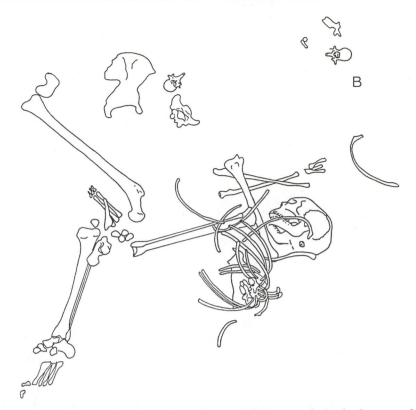

Figure 5. The Neolithic human bog sacrifices from Sigersdal, Denmark, include two adolescents, probably females, around 16 and 18-20 years old, found together with a lugged flask and animal bones, mainly from cattle. The best preserved body seen above had a large lesion on the left side of the skull and a rope around the neck. The skeletons are ^{14}C dated to 2700 b.c. (Courtesy of Pia Bennike and Klaus Ebbesen 1987.)

years. The initial contact with farming communities to the south took place around 3600 b.c., domesticates were introduced to Scandinavia around 3100 b.c., but the first major agricultural expansion did not happen before 2600 b.c. This transition can be viewed in terms of four stages within the Ertebølle and the Tragtbæger periods: Mesolithic, Last Hunters, First Farmers and Neolithic (Figure 2). These stages describe (1) the Mesolithic in southern Scandinavia, (2) the time of first contact with continental farmers during the late Mesolithic, (3) the period of initial farming in the earliest Neolithic, and (4) the later Tragtbæger with a fully established neolithic economy.

Mesolithic

The first part of the Ertebølle period represents a time of pristine foraging communities. There is no evidence of any contact with farming groups in central and southern Europe. These foragers in southern Scandinavia lived in sedentary communities in both inland and coastal areas. Subsistence was broadly based on a large number of species. In the coastal areas, dependence on marine resources such as fish and

mammals was high. An elaborate technology of wood, bone, antler, fabric, and stone materials were in use. Population was relatively high in number, particularly in coastal areas, and evidence of territorial behavior is reflected in the distribution of artifacts and styles and in signs of inter-group conflict and violent death. The dead were buried in cemeteries which reflect differences by age and sex, but provide no indication of status differentiation among the members of society.

Last Hunters

Shortly after 3600 b.c., Ertebølle foragers began to come into contact with farming populations to the south. By this time, descendants of Linearbandkeramik farmers had expanded from Hungary across Central Europe and into northern Germany and southern Poland, a distance of 100 km or so from the coastal Ertebølle groups. Ideas as well as exotic items were imported to southern Scandinavia, the adoption of pottery technology being perhaps the most important innovation. A number of other "Neolithic" items were borrowed and appeared in a Mesolithic context, including T-shaped antler axes, bone rings, and bone combs (Andersen 1980, Vang Petersen 1984). The presence of the shaft-hole axes (*Schuhleistenkeile*), made from a special type of amphibolite found only in southeastern Europe, provides evidence of long distance import from farming communities to the south (Fischer 1982).

Despite this exposure to the skills and material culture of neolithic communities, it appears that other aspects of these farming societies were of little interest to Scandinavian hunter-gatherers. Evidence of cereal and domesticated animals is virtually absent and there is no indication that the last hunters in Scandinavia changed their basic pattern of subsistence and settlement before around 3100 b.c.— at least 500 years after the initial contact with southern farmers.

First Farmers

Domesticated cereals and animals were introduced to Southern Scandinavia around 3100 b.c. along with polished flint axes, a thin-walled type of pottery and new pot shapes (TRB pottery), grinding stones, and the ard. New weapons and male status symbols appear at the same time together with amber and copper jewelery. However, the technology of pottery making and the unpolished flint industry is derived from the Ertebølle.

The settlement pattern is one of small residential sites combined with seasonal hunting camps suggesting that the first farmers lived in fairly small and somewhat mobile groups. Both settlement patterns and carbon isotope analysis indicate a shift in diet toward a greater reliance on terrestrial food. Animal husbandry appears to be of primary importance during this period; only limited amounts of cereal seem to have been grown in this stage. Hunting, fishing, and gathering obviously remained an important supplement to the economy.

New social and religious patterns emerged with the first farmers. An increasing amount of surplus production was invested in rituals related to ancestor cult and perhaps fertility cult. Large burial monuments were erected for a few individuals; sacrifices of food, domesticated animals, and humans were made in wetland areas (Figure 5), sometimes with distinct evidence for feasting. Intensive trade and exchange promoted the movement of prestige items and utilitarian commodities inside Scandinavia and provided contact with central Europe.

Neolithic

It is not until 2600 b.c.—500 years after the introduction of domesticates—that major agricultural expansion took place in southern Scandinavia. Vast areas of forest were cleared, and there is significant evidence of cattle herding and the use of pasture as well as cereal cultivation, predominantly of wheat.

Settlements increased dramatically in size and number. Substantial houses were constructed. Territorial divisions appeared to be fixed; each hamlet was marked by a cluster of megalithic tombs, and a group of hamlets shared a common regional ceremonial center at the causewayed enclosures. An enormous amount of energy was invested in ancestor cult and other rituals. The majority of the thousands of megalithic tombs in Southern Scandinavia were constructed during the fairly short period between 2800–2450 b.c. Huge celebrations at the causewayed enclosures would have required the participation of several hundred people. Trade and exchange of flint axes, copper, and amber items was intensive at this time. The heavier reliance on food production apparently was associated with a great need for more rituals and for status symbols to support rival demands for power and control of people and resources.

During the late TRB, between 2450–2200 b.c., large villages dependent on cereal growing and cattle herding are found at close intervals in an open landscape. Rituals were reduced to a minimum and old tombs were reused rather than new ones constructed. The amount and intensity of trade and exchange was significantly reduced. After a period of exorbitant investments in rituals, farming had definitely become the settled and stable way of life.

Theoretical Perspectives on the Transition

The explanation of the transition to agriculture in Scandinavia has been and still is the subject of a long and fervent discussion. Prior to the 1970s the primary topic of debate was whether the transition represented a colonization by TRB people or a transformation of the indigenous Ertebølle into an agrarian society. The invading TRB people were thought by some to derive from either the western or the eastern part of Central Europe, or even from the Ukraine (Becker 1948, Lichardus 1976). Others argued that the TRB represented an independent Scandinavian development from the local Ertebølle (Troels-Smith 1953). Today most archaeologists believe the TRB developed from the local Mesolithic under greater or lesser influence from various Danubian groups (Andersen 1975, Fischer 1982, Jennbert 1984, 1985, Jensen 1979, Madsen and Petersen 1984, Nielsen 1985, 1987, Paludan-Müller 1978, Rowley-Conwy 1985, Zvelebil and Rowley-Conwy 1984, 1986). These discussions involve consideration of whether TRB emerged from a single source (Nielsen 1985) or if the regional variation within the early TRB represents several independent centers of development (Madsen and Petersen 1984). Simple colonization and replacement of the Mesolithic population is ruled out by the continuity in flint and pottery technology.

Most recent theories about why Ertebølle foragers opted for agriculture involve either the stress resulting from changes in population or environment, or demands from increasing social prestige. A number of opposing explanations have been suggested, ranging from a combination of small-scale migration and indigenous adoption by late Mesolithic groups (Larsson 1987), to climate change (Larsson 1987, Rowley-Conwy 1985), the gradual spread of ideas and products (Nielsen 1985), resource-poor inland groups seeking a productive source of food (Madsen 1986), or exchange cycles and competition for prestige (Fischer 1982, Jennbert 1984, 1985). The paragraphs below outline this debate.

Some imbalance between population levels and available resources is implied in a number of papers on the question of the transition (Andersen 1975, 1981, Larsson 1987, Rowley-Conwy 1983, 1984, 1985, Zwelebil and Rowley-Conwy 1984, Paludan-Müller 1978). Larsson (1987), for example, considers environmental changes and small-scale immigration of Neolithic farmers as the explanation for the introduction of Neolithic elements and their rapid spread over all of southern Scandinavia. According to Larsson (1987) changes in the coastal environment, due to the retreat of the sea and a decline in temperature, diminished the role of marine resources and increased the exploitation of inland plants and animals. Rowley-Conwy (1985) argues that the transition was a result of changing climate at the end of the Atlantic climatic episode. Lower salinity as a result of higher ocean temperature may have been responsible for a demise of the shellfish beds, an important supplement to the diet. This loss required new sources of food for the existing population, and farming was available nearby to supply that need. However, shellfish were exploited only in certain areas of southern Scandinavia and their local demise cannot explain the widespread appearance of cultigens and TRB at 3100 b.c.

Other scholars suggest that the transition was caused not by environmental constraints, but rather by factors involving social structure and the emergence of status inequality. Fischer (1982) and Jennbert (1984, 1985) have argued for close connections between the farmers of north-central Europe and the foragers of Denmark, pointing to a number of "borrowed" artifacts and ideas, including ceramics and certain stone and antler axes in the late Mesolithic. They suggest that these successful foragers did not require additional sources of food—that the only obvious reason for farming was to generate surplus. Jennbert suggests that certain leaders were likely responsible for encouraging the accumulation of wealth through cultivation

and herding. Competition between higher status individuals for prestige then might explain why successful foragers adopted farming.

Important Variables in the Transition to Farming

This section considers the transition to agriculture in terms of several potentially important variables in current theories regarding the adoption of farming, including environmental change, population size and density, circumscription, resource availability and stress, and social differentiation, and concludes with a discussion of the rate of change of the transition.

Climatic and Environmental Change

Discussions regarding the role of the environment in the transition to agriculture in southern Scandinavia generally involve change in climate from Atlantic to Subboreal. The discussion has focused particularly on a decline in marine resources due to changes in sea level, ocean temperature, or water salinity. These changes are assumed to upset the balance between population and available resources and cause a situation of shortage where the adoption of farming becomes a necessary response (Andersen 1973, 1981, Larsson 1987, Rowley-Conwy 1983, 1984, 1985, Zwelebil and Rowley-Conwy 1984, Paludan-Müller 1978).

The shift from Atlantic to Subboreal conditions occurred ca. 3000 b.c., at the same time of the transition to agriculture. During the early Atlantic period rising sea levels caused the formation of the North Sea and turned the western part of the Baltic Sea into a saltwater ocean. These changes in sea level greatly reduced the land surface available to human populations in northwestern Europe. Oscillating transgressions and regressions continued throughout the Atlantic and Subboreal period, but these later changes in sea levels were less dramatic. The bays, inlets, and estuaries created by rising sea levels were among the richest resource zones for Mesolithic groups. In general the Atlantic climate was moist and warm, promoting the formation of a dense, mature forest. Earlier interpretations of this forest as dark, monotonous, and increasingly poor in game appear to be exaggerated. The disappearance of a few species such as elk and aurochs during the Atlantic is likely related to island formation, isolation, and extirpation by human hunting rather than the closing of the forest; game populations in general appear to be rich (Aaris-Sørensen 1988).

The Subboreal was slightly cooler and drier than the preceding Atlantic. A decrease in the pollen of ivy indicates cooler winters when the fjords and inlets might be frozen. Even though winter temperatures may have been slightly lower, annual averages remained higher than today. Simulations of past climate (e.g., Kutzbach and Guetter 1986) also suggest that conditions in southern Scandinavia at 5000 b.c. would have been slightly warmer than present, particularly in the summer. The presence of mistletoe and a certain species of tortoise in the Subboreal, both found only to the south of Denmark today, suggests that environmental conditions after 3000 b.c. were still quite favorable. The environmental changes associated with the shift from the Atlantic to the Subboreal were not major. The Littorina Sea continued to be rich in fish and sea mammals, and a considerable game population inhabited the forest. It does not seem possible to argue for an ecological crisis at the time of the adoption of farming.

The elm decline in the pollen record is also an indicator of the shift from Atlantic to Subboreal climate or from older to younger Lime forest period. The simultaneous appearance of the first evidence of farming and the elm decline originally gave support to the assumption that the elm decline was caused by human land clearance and related to the harvesting of leaf fodder for livestock (Rasmussen 1990). It is, however, unlikely that humans alone could be the cause of an environmental disaster of this magnitude. Alternative explanations of the elm decline include elm disease, soil deterioration, and climatic change perhaps fostered by the activities of man (Berglund 1985, Kolstrup 1987, Rasmussen 1990).

Resource Availability and Food Stress

What is perhaps most important to remember about the environment of early Holocene northern Europe is its *diversity*. A rich mix of plants and animals occupied this landscape and the surrounding oceans, providing a wealth of resources to the human occupants. The late Atlantic was perhaps the most favorable period for hunter-gatherers with temperatures warmer than today and stable climate and vegetation. Abundant terrestrial and marine resources were available providing a number of options to fulfill basic needs for food and clothing. The number of specialized hunting sites for fish, seals, and fowl suggest that Mesolithic hunters were quite adept at exploiting these options. The use of marine animals by coastal populations meant that over-exploitation of these food sources was very unlikely. Inland Ertebølle populations, however, may have pursued a pattern of economic exploitation and a degree of sedentism that extended the capacity of the environment (Madsen 1986).

The same resources generally were available during the Subboreal period. The diet of the first farmers changed toward a much greater reliance on terrestrial foods. The role of domesticates, cereal in particular, appears to have been minor for several hundred years; wild food sources remained an important component in the subsistence economy. The intensification of food production happened gradually and there is no evidence for a major agricultural expansion before 2600 b.c.—500 years after the introduction of domesticates. The varied subsistence activities of the first farmers and the relative unimportance of domesticates at that time do not suggest that domesticates were required to alleviate food shortages.

The available biological evidence from human skeletons in the Mesolithic and Neolithic also indicates that food stress was not involved in the transition from foraging to farming. Mesolithic skeletons are generally robust with little indication of disease or malnutrition (Frayer 1980, Meiklejohn et al. 1984). In contrast, Neolithic skeletal remains generally evidence more dental disease and smaller overall size.

Population Size and Density

An increase in numbers or density of population toward the end of the Mesolithic might make necessary some means for increasing the amount of food available. However, population numbers and density are notoriously difficult to extract from archaeological data and only a few general comments can be made, along with a summary of some preliminary survey results.

The systematic survey of the Saltbæk Vig (Gebauer and Price 1990) provides some estimate of the number and size of sites from each period of concern. The survey covered approximately 3.26 km² of agricultural fields along the coast of western Zealand and resulted in the location of approximately 150 sites, including single finds and settlements. The distribution of these sites is shown in Figure 6. Chronological designation of these find spots was based on diagnostic artifacts types. Approximately 25 of the sites belong to the Mesolithic period, 97 are Neolithic in age, and the remainder are unknown or mixed. A total of 15,618 artifacts were collected and counted during the survey. The largest single site contained 1306 pieces while the smallest sites were single finds. The average size of a surface collection from the Mesolithic sites was 99.1 artifacts; the average size of a collection from the Neolithic sites was 108.9. The largest Mesolithic collection contained 494 pieces; the largest Neolithic site contained 1306 pieces. Distribution of the collection sizes by period (Figure 6) indicates the general similarity of the materials, but there is more diversity among the Neolithic sites,

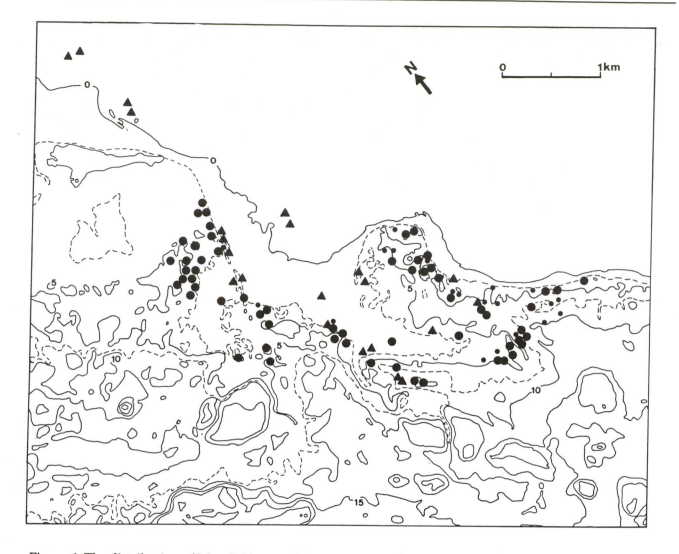

Figure 6. The distribution of Mesolithic sites (triangles), Neolithic sites (large circles), and single finds (small circles) in the Saltbæk Vig survey area (Gebauer and Price 1990).

including both smaller and larger sites. In terms of the size of the collection area, Mesolithic sites averaged 2447 m² while Neolithic sites averaged 2239 m².

Survey information from the Saltbæk Vig area shows a fairly even distribution among Mesolithic and Neolithic sites. Neolithic sites tend to be richer in terms of artifact density and more diverse in settlement size, whereas some of the Mesolithic sites surpass the largest Neolithic sites in actual size. This pattern does not suggest any dramatic changes in population density, but may indicate a trend toward a larger population in the Neolithic. However, this information provides only a broad comparison between Mesolithic and Neolithic; it is not possible to distinguish sites belonging to the period *immediately* before and after

the transition to agriculture.

Similar survey information elsewhere indicates an intensification in subsistence and settlement through the Mesolithic with sedentary communities present during the Ertebølle period. Apparently there was a relatively dense population during the Ertebølle, especially at the coast; however, there is no obvious evidence for population pressure preceding the transition to agriculture. The first farmers were settled in small scattered habitations combined with seasonal extraction camps. Population appears to increase somewhat around 2800–2900 b.c., and in particular around 2600 b.c., but there is no evidence of increasing population size immediately following the introduction of agriculture at 3100 b.c.

Circumscription and Territorial Behavior

Evidence from the distribution of various artifacts and materials in northern Europe suggests distinctive zones. While it is not yet clear what such zones mean in terms of specific social organization, a general pattern of a reduction in territory over time seems apparent. In southern Scandinavia, a variety of artifacts, materials, and designs show restricted distributions particularly during the latter part of the Mesolithic.

Regional variation in artifact types and styles is documented from the later half of the Ertebølle between eastern and western Denmark. Groundstone Limhamn axes occur largely on Zealand and in Scania (Becker 1939), while T-shaped antler axes and bone combs and rings are found primarily in Jutland (Andersen 1975). Small groupings at the local level can be separated on the basis of certain design elements on pottery (Andersen 1975). Analysis of flaked flint axes from Zealand (Vang Petersen 1984) has indicated that certain types are distinctive to local areas within the northeast corner of the island, covering areas with a diameter on the order of 40 km or less. The shift from larger areas of 100,000 km^2 to areas of 1000 km^2 or even smaller is an important indication of the increasing density and definition of human groups in this period (Price 1981, Verhart 1990).

Increased territorial behavior during the Ertebølle is also reflected in the violent trauma found not uncommonly in the human skeletal population. A number of the individuals from Mesolithic cemeteries appear to have been victims of murder (Bennike 1985, Persson and Persson 1984). The existence of these cemeteries or "formal disposal areas" is further indication of a strong territorial behavior in the Ertebølle society (Chapman 1981, Larsson 1990). Most likely the cemeteries reflect the presence of some form of corporate group.

Apparently a large permanent coastal population was settled within well defined territories that were maintained with a good bit of conflict. Dietary differences suggest a specialization between coastal and inland Ertebølle populations (Noe-Nygaard 1988). Information on the life style of the inland populations is limited, in large part because of factors of preservation. However, this dichotomy of Ertebølle settlement adds to the picture of circumscription promoted by economic specialization and a desire to control localized resources.

Neither settlement pattern nor regional variation in the distribution of artifacts or styles reflect the same degree of circumscription during the initial period of farming. A similar distinct pattern of local groupings within broader regional zones does not appear again before around 2600 b.c. or perhaps a little later. However, the construction of monumental earthen long barrows in the earliest Neolithic is likely an expression of territorial rights by the local inhabitants, symbolically legitimized through their ancestors' relations to the land. Local ceremonial sites in wetland areas may be further indication of the relationship between the local community and a certain territory. A higher level of organization is not apparent before the construction of regional ceremonial centers at causewayed enclosures beginning around 2600 b.c.

Social Differentiation

Hypotheses invoking social causes for the transition to agriculture suggest that the adoption of domesticates took place in a situation of emerging social inequality. Higher status individuals may have encouraged the production of food surpluses to be used at competitive feasts or in exchange for prestige items. Domesticated foods were especially valuable in this context as they were new and exotic and could be stored and accumulated. Some kinds of domesticates were particularly suited for feasting; the cultivation of cereals, for example, may initially have been related primarily to the brewing of beer (Braidwood 1953, Katz and Maytag 1991).

No evidence of elite burial treatment is found in the Ertebølle. The organization and maintenance of long distance exchange networks does, however, suggest some kind of status differentiation, whether related to individuals or groups. Prestige items like the Neolithic shaft-hole axes (*Schuhleistenkeile*) may have served as status symbols for certain males in the Ertebølle as well. A certain amount of surplus production was clearly invested in maintenance of exchange relationships, but similar evidence of surplus accumulation is not found in other spheres of society. Competition for control of people and resources is reflected in the overt territorial behavior. Evidence of domesticates are very scarce though and the few examples found in the Ertebølle may reflect a "Fertile Gift" obtained through exchange rather than local production (Jennbert 1984).

Territorial behavior appears to be less pronounced among the first farmers. Only broad regional variation is found in artifact styles and pottery decoration. More substantial evidence is found of status differentiation than during the Ertebølle, however, particularly in the elite burials in earthen long barrows. Status symbols for the elite are also seen in exotic copper jewelery, copper axes, the relatively rare, stone battle axes, and flint axes of extraordinary size and beauty. An intensive exchange system circulated prestige items as well as other commodities. Sacrifices of food, domesticated animals and humans took place at wetland areas, probably related to fertility rituals. Distinct evidence of feasting is found at some of these sites.

New social and religious patterns apparently were established at the same time domesticates were introduced in Scandinavia. The fierce competition for resource control among the last hunters appears to have been transformed into rituals related to ancestor and fertility cults among the first farmers. Food and drinks were obviously an important part of these rituals. Clay flasks or little beakers indicate that some kind of drink accompanied the dead in the afterlife (Figure 7). Domesticated animals, especially cattle, and pots presumably filled with food form the main part of the sacrifices at wetland areas. Domesticated plants and animals may have been an important component in the rituals and indirectly in the negotiation of social and political relationships among the first farmers. The Scandinavian data may support the hypothesis that food production was adopted as a means of accumulating surplus and gaining power through acquisition of prestige items and feasting. However the relationship between emerging social complexity and the adoption of farming is still a chicken-egg argument.

Rate of Change

The rate of change in the transition to agriculture in southern Scandinavia is very informative with regard to how this process occurred. There are, however, two responses to questions about the rate of change—it is both fast and slow, depending upon the specific aspects under consideration.

On the one hand, the period from first contact with farmers until the full adoption of agriculture extends over more than 1000 years, from around 3600 b.c. until 2600 b.c. Sedentary foragers in the late Mesolithic began to import items of Neolithic manufacture from the south after 3600 b.c. The appearance of TRB pottery, domesticates, and long barrows marks the recognized beginning of the Neolithic around 3100 b.c., but a fully Neolithic economy is not in place until after 2600 b.c. On the other hand, the almost simultaneous appearance of the evidence for domesticated plants and animals, TRB pottery, and earthen long barrows across southern Scandinavia around 3100 b.c. is remarkable (Madsen n.d.). These two phenomena are not immediately reconcilable.

Thus, the nature of the transition is not clear. Was it a rapid change instantly transforming Ertebølle hunter-gatherers into Tragtbæger farmers? Or was it a very gradual element by element replacement of the Ertebølle (Jennbert 1984). According to the latter approach, increasing social complexity within Ertebølle society combined with the presence of cultigens made available through exchange promoted by the step-by-step transition to farming. Evidence of a gradual replacement of Ertebølle elements with

Tragtbæger forms is found in the stratigraphy at the site of Löddesborg (Jennbert 1984). The continuous use of this large settlement through the transition suggests a high degree of cultural continuity between the Ertebølle and the Tragtbæger. Even if the integrity of the stratigraphy at this site has been questioned (Nielsen 1987), the presence of grain impressions in the pottery clearly indicates that domestic cereals were known to the Ertebølle inhabitants. A few pollen diagrams with evidence for the presence of pasture prior to the elm decline also suggest that domesticated animals may have been imported during the late Mesolithic.

Most archaeologists perceive the transition as a thorough and rapid transformation based upon the simultaneous appearance of domesticates and Tragtbæger characteristics all over southern Scandinavia (Madsen 1986, Nielsen 1987). Change in settlement pattern, distinctly different resource exploitation at the kitchen middens, change in diet, change in rituals and perhaps religious beliefs all together suggest that pervasive alterations of the society took place at 3100 b.c. The magnitude and abruptness of this transition can hardly be questioned. Even so, the very long period of contact prior to the use of cultigens and the very slow development toward full reliance on agriculture remains a puzzle.

Conclusions

Evaluation of the hypotheses and important variables concerned with the transition to agriculture in southern Scandinavia is a difficult process. The critical variables of population and resource availability that are employed in a number of models of the transition are very difficult to retrieve archaeologically. Perhaps one of the reasons that these variables continue to be invoked and debated lies in their high level of immeasurability. Explanations concerned with changes in society are also intriguing, and also difficult to evaluate. Certainly parts of the Danish Neolithic data fit such a model nicely, including evidence for feasting at bog sacrifice sites and later at causewayed enclosures, the burial of single individuals in the first large tombs, the variety of trade items in circulation, and perhaps even the adoption of domestic plants and animals. Nevertheless, suggestions of social inequality, competitive feasting, and the emergence of big men as causes for the adoption of agriculture remain to be demonstrated. It is difficult to determine if the accumulation of surplus was a cause or a consequence of emerging social inequality.

Clearly, the evidence and the arguments concerning the shift to agriculture in southern Scandinavia are complex and not yet readily resolvable. There is a

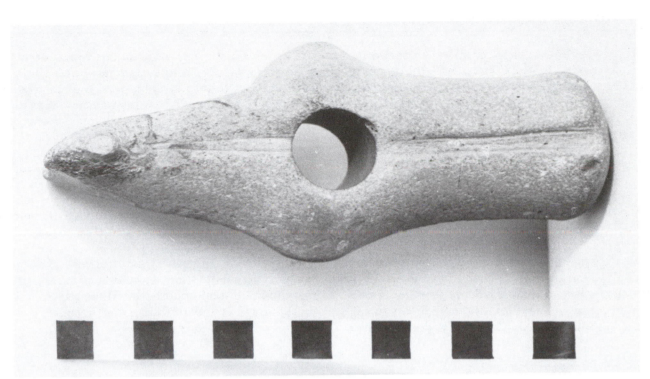

Figure 7. Votive offerings from an elite burial in the Rustrup long barrow in east-central Jutland included a small TRB pot, approximately 10 cm high, and a polygonal battle axe. The burial is ^{14}C dated to 3020 b.c. (© Silkeborg Museum, photos by Lars Bay.)

poor fit between the archaeological evidence and the explanatory models in use. Concern with the origins of TRB has been largely over "how," not "why," the transition occurred. As a consequence of perceived differences between the Mesolithic and the Neolithic, studies have often focused on one side of the transition or the other, rather than viewing it as a continuous process. New views of the late Mesolithic and early Neolithic have not clarified the nature of the transition between them. It seems essential to consider the process of the transition across the Mesolithic-Neolithic boundary, rather than simply comparing the two periods as largely different entities.

Several points deserve reiteration in conclusion. First of all, the transition to agriculture in northern Europe takes place in an area where successful, sedentary hunter-gatherers apparently have plenty to eat. This is not a marginal zone. Secondly, this is an area

with a long history of archaeological research and good preservation. The density of archaeological data is high. It is generally clear who and what were involved, and when and how the transition took place. Third, in spite of the quality and volume of evidence, the question of *why* humans adopted agriculture remains elusive. It is not yet possible to document unequivocally the direct and specific factors involved in the transition in this area. A number of major questions remain to be answered and existing evidence needs to be confirmed, regarding both the late Mesolithic and the early Neolithic in southern Scandinavia. An absence of solution, in the face of good evidence, suggests that either we need to be asking different questions or getting even more information. In all likelihood both will be required for a breakthrough in understanding this process.

References Cited

Aaris-Sørensen, K.
1988 *Danmarks forhistoriske Dyreverden*. Fra Istid til Vikingetid. Copenhagen.
Albrethsen, S. E., and E. Brinch Petersen
1976 Excavation of a Mesolithic Cemetery at Vedbæk, Denmark. *Acta Archaeologica* 47:1-28.
Andersen, N.H.
1981 Sarup. Befæstede neolitiske anlæg og deres baggrund. *Kuml* 1981:63-103.
1988 The Neolithic Causewayed Enclosures at Sarup, on South-West Funen, Denmark. In *Enclosures and Defences in the Neolithic of Western Europe*, edited by C. Burgess et al., pp 337-363. BAR International Series 403. BAR, Oxford.
Andersen, S.H.
1975 Ringkloster, en jysk indlandsboplads med Ertebøllekultur. *Kuml* 1973-74: 11-108.
1976 Et østjysk fjordsystems bebyggelse i stenalderen. *Bebyggelsesarkæologi*, edited by H. Thrane, pp. 18-61. Skrifter fra Historisk Institut, Odense Universitet 17.
1980 Ertebøllekunst. Nye østjyske fund af mønstrede Ertebølleoldsager. *Kuml* 1980:7-62.
1981 *Danmarkshistorien. Jægerstenalderen*. Sesam, Copenhagen.
Andersen, S. H. and E. Johansen.
1987 Ertebølle revisited. *Journal of Danish Archaeology* 5. 1986:31-61.
Becker, C.J.
1939 En stenalderboplads paa Ordrup Næs i Nordvestsjælland. Bidrag til Spørgsmaalet om Ertebøllekulturens varighed. *Aarbøger for Nordisk Oldkyndighed og Historie* 1939:199-280.
1948 Mosefundne lerkar fra Yngre Stenalder. Studier over yngre stenalder i Danmark. *Aarbøger for Nordisk Oldkyndighed og Historie* 1947:1-318.
Bennike, P.
1985 *Palaeopathology of Danish Skeletons. A Comparative Study of Demography, Disease and Injury*. Akademisk Forlag, København.
Bennike, P. and K. Ebbesen
1987 The Bog Find from Sigersdal. Human Sacrifice in the Early Neolithic. *Journal of Danish Archaeology* 5: 83-115.
Berglund, B. E.
1985 Early Agriculture in Scandinavia: Research Problems related to Pollen-analytical Studies. *Norwegian Archaeological Review* 18:77-105.

Blankholm, H.P.
 1987 Late Mesolithic Hunter-gatherers and the Transition to Agriculture in Southern Scandinavia. In *Mesolithic Northwest Europe: Recent Trends*, edited by P. Rowley-Conwy, M. Zvelebil, and H.P. Blankholm, pp. 155-162. University of Sheffield, England.

Braidwood, R.J.
 1953 Symposium: Did Man Once Live by Beer Alone? *American Anthropologist* 55: 515-526.

Chapman, R.
 1981 The Emergence of Formal Disposal Areas and the 'Problem' of Megalithic Tombs in Prehistoric Europe. *The Archaeology of Death*, edited by R. Chapman, I. Kinnes, and K. Randsborg, pp. 71-81. Cambridge University Press, Cambridge.

Ebbesen, K.
 1975 *Die jungere Trichterbecherkultur auf den Dänischen Inseln.* Arkæologiske Studier, Copenhagen.
 1978 *Tragtbægerkulturen i Nordjylland.* Nordiske Fortidsminder, Ser. B 5, København.
 1979 *Stordyssen i Vedsted. Studier over den sønderjyske Tragtbægerkultur.* Arkæologiske Studier VI, København.
 1982 Yngre Stenalders Depotfund som bebyggelseshistorisk kildemateriale. In *Om yngre stenalders bebyggelseshistorie*, edited by H. Thrane, pp. 60-79. Skrifter fra Historisk Institut, nr. 30 Odense universitet. Odense.

Fischer, A.
 1982 Trade in Danubian Shaft-Hole Axes and the Introduction of Neolithic Economy in Denmark. *Journal of Danish Archaeology* 1:7-12.

Frayer, D.W.
 1980 Sexual Dimorphism and Cultural Evolution in the Late Pleistocene and Holocene of Europe. *Journal of Human Evolution* 9:399-415.

Gebauer, Anne Birgitte
 1988a Stylistic Variation in the Pottery of the Funnelbeaker Culture. In *Multivariate Archaeology. Numerical Approaches in Scandinavian Archaeology*, edited by T. Madsen, pp. 91-117. Jutland Archaeological Society Publications.
 1988b The Long Dolmen at Asnæs Forskov, West Zealand. *Journal of Danish Archaeology* 7:40-52.

Gebauer, Anne Birgitte, and T. Douglas Price
 1990 The End of the Mesolithic in Eastern Denmark: A Preliminary Report on the Saltbæk Vig Project. In *Contributions to the Mesolithic in Europe*, edited by P.M. Vermeersch and P. van Peer, pp. 259-280. Studia Praehistoric Belgica 5, Leuven University Press.

Helbæk, H.
 1955 Store Valby—Kornavl i Danmarks første neolitiske fase. *Aarbøger for Nordisk Oldkyndighed og Historie* 1954:198-204.

Hjelmquist, H.
 1979 Beiträge zur Kenntnis der Prähistorischen Nutzpflanzen in Schweden. *Opera Botanica* 47:1-58. Stockholm

Iversen, J.
 1941 Landnam i Danmarks stenalder. *Danmarks Geologiske undersøgelser*, II Række, Nr. 66.

Jennbert, K.
 1984 *Den produktiva gåvan. Tradition och innovation i Sydskandinavien för omkring 5300 år sedan.* Acta Archaeologica Lundensia, Series in 4, No. 16. CWK Gleerup, Lund.
 1985 Neolithisation—a Scanian Perspective. *Journal of Danish Archaeology* 4:196-197.

Jørgensen, G.
 1977 Et kornfund fra Sarup. Bidrag til Belysning af Tragtbægerkulturens Agerbrug. *Kuml* 1976:47-64.

Katz, S. H., and F. Maytag
 1991 Brewing an Ancient Beer. *Archaeology* July 1991.

Kolstrup, E.
 1987 Tidligt landbrug. *Skalk* 5:9-12.

Kutzbach, J.E. , and P.J. Guetter
 1986 The Influence of Changing Orbital Parameters and Surface Boundary Conditions on Climate Simulations for the Past 18,000 years. *Journal of Atmospheric Sciences* 43:1726-1759.

Larsson, L.
  1987 Some Aspects of Cultural Relationship and Ecological Conditions during the Late Mesolithic and Early Neolithic. In *Theoretical Approaches to Artefacts, Settlement and Society. Studies in Honour of Mats P. Malmer*, edited by G. Burenhult et al., pp. 165-176. BAR International Series 366. BAR, Oxford.
  1988 *The Skateholm Project. I.* Acta Regiae Societatis Humaniorum Literarum Ludensis, LXXIX. Almqvist & Wiksell, Stockholm.
  1990 The Mesolithic of Southern Scandinavia. *Journal of World Prehistory* 4: 257-310.
Larsson, L., and M. Larsson
  1986 Stenåldersbebyggelse i Ystadsområdet. *Ystadiana* 1986.
Larsson, M.
  1986 Neolithisation in Scania—A Funnel Beaker Perspective. *Journal of Danish Archaeology* 5:244-247.
Lichardus, J.
  1976 *Rössen - Gatersleben - Baalberge*. Ein Beitrag zur Chronologie des Mitteldeutschen Neolitikums und zur Entstehung der Trichterbecher-Kulturen. Bonn.
Madsen, T.
  1979 Earthen Long Barrows and Timber Structures: Aspects of the Early Neolithic Mortuary Practice in Denmark. *Proceedings of the Prehistoric Society* 45:301-320.
  1982 Settlement Systems of Early Agricultural Societies in East Jutland, Denmark: A Regional Study of Change. *Journal of Anthropological Archaeology* 1:197-236.
  1986 Where did all the Hunters go? *Journal of Danish Archaeology* 5:230-239.
  1989 Causewayed Enclosures in South Scandinavia. Forthcoming.
Madsen, T., and H. Juel Jensen
  1982 Settlement and land use in Early Neolithic Denmark. *Analecta Praehistorica Leidensia* 15:63-86.
Madsen, T., and J. E. Petersen
  1984 Tidligneolitiske anlæg ved Mosegården. Regionale og kronologiske forskelle i tidligneolitikum. *Kuml* 1982-83:61-120.
Meiklejohn, C., C. Schentag, A. Venema, and P. Key
  1984 Socioeconomic Change and Patterns of Pathology and Variation in the Mesolithic and Neolithic of Western Europe: Some Suggestions. In *Paleopathology at the Origins of Agriculture*, edited by Cohen and Armelagos, pp. 75-100. Orlando, Academic Press.
Nielsen, E. K.
  1983 *Tidligneolitiske keramikfund*. Unpublished Ph.D. thesis, Copenhagen University.
  1987 Ertebølle and Funnelbeaker Pots as Tools. On Traces of Production Techniques and Use. *Acta Archaeologica 1986* 57:107-120.
Nielsen, P.O.
  1977 Die Flintbeile der frühen Trichterbecherkultur in Dänemark. *Acta Archaeologica* 48:61-138.
  1985 De første bønder. Nye fund fra den tidligste Tragtbægerkultur ved Sigersted. *Aarbøger for Nordisk Oldkyndighed og Historie* 1984:96-126.
  1986 The Beginning of the Neolithic—Assimilation or Complex Change? *Journal of Danish Archaeology* 5:240-243.
Noe-Nygaard, N.
  1983 The Importance of Aquatic Resources to Mesolithic Man at Inland Sites in Denmark. In *Animals and Archaeology 2. Shell Middens, Fishes and Birds*, edited by C. Grigson and J. Clutton-Brock, pp.125-142. BAR International Series 183. BAR, Oxford.
  1988 δ^{13}C-values of Dog Bones Reveal the Nature of Changes in Man's Food Resources at the Mesolithic-Neolithic Transition in Denmark. *Chemical Geology* 73:87-96.
Paludan-Müller, C.
  1978 High Atlantic Food Gathering in Northwest Zealand, Ecological Conditions and Spatial Representation. In *New Directions in Scandinavian Archaeology, vol. 1*, edited by K. Kristiansen and C. Paludan-Müller, pp. 120-157. Studies in Scandinavian Prehistory and Early History, Odense.
Persson, O., and E. Persson
  1984 Anthropological Report on the Mesolithic Graves from Skateholm, Southern Sweden I. Excavation Seasons 1980-1982. *Report Series No. 21*. University of Lund Institute of Archaeology, Lund.

Price, T. Douglas.
  1981 Regional Approaches to Human Adaptation in the Mesolithic of the North European Plain. In *Mesolithikum in Europa*, edited by B. Gramsch, pp. 217—234. Veröffentlichungen des Museums für Ur- und Frühgeschicht Potsdam 14/15.
  1985 Affluent Foragers in Mesolithic South Scandinavia. In *Prehistoric Hunter-Gatherers: The Emergence of Cultural Complexity*, edited by T.D. Price and J.A. Brown, pp. 341-363. Academic Press, Orlando.
  1986 The Earlier Stone Age of Northern Europe. In *The End of the Paleolithic in the Old World*, edited by L.G. Strauss, pp. 1-30. BAR International Series S284. BAR, Oxford.
  1987 The Mesolithic of Western Europe. *Journal of World Prehistory* 1:225-332.
  1989 The Reconstruction of Mesolithic Diets. In *The Mesolithic in Europe: Proceedings of the Third International Symposium*, Edinburgh 1985, edited by C. Bonsall, pp. 48-59. University Press, Edinburgh.

Randsborg, K.
  1979 Resource Distribution and the Function of Copper in Early Neolithic Denmark. In *The Origins of Metallurgy in Atlantic Europe*, edited by M. Ryan, pp. 303-318. Proceedings of the Fifth Atlantic Colloquium. Stationary Office, Dublin.

Rasmussen, P.
  1990 Leaf Foddering in the Earliest Neolithic Agriculture: Evidence from Switzerland and Denmark. *Acta Archaeologica* 60: 71-86.

Rech, M.
  1979 Studien zu Depotfunden der Trichterbecher- und Einzelgrabkultur des Nordens. *Offa-Bücher* 39. Neumünster.

Rowley-Conwy, Peter.
  1983 Sedentary Hunters: the Ertebølle Example. In *Hunter-gatherer Economy*, edited by G. Bailey, pp. 111-126. Cambridge University Press.
  1984 The Laziness of the Short-Distance Hunter: The Origin of Agriculture in Western Denmark. *Journal of Anthropological Archaeology* 3:300-324.
  1985 The Origin of Agriculture in Denmark: A Review of Some Theories. *Journal of Danish Archaeology* 4:188-195.

Runnels, C., and T.H. van Andel
  1988 Trade and the Origins of Agriculture in the Eastern Mediterranean. *Journal of Mediterranean Archaeology* 1:83-109.

Sherratt, A.
  1976 Resources, Technology, and Trade: an Essay in Early European Metallurgy. In *Problems in Economic and Social Archaeology*, edited by G. Sieveking, I. Longworth, and K. Wilson, pp. 557-582. Gerald Duckworth & Co., London.

Skaarup, J.
  1973 *Hesselø-Sølager. Jagdstationen der südskandinavischer Trichterbecherkultur.* Arkæologiske Studier 1. Akademisk Forlag, København.
  1975 *Stengade. Ein langeländischer Wohnplatz mit Hausresten aus der frühneolitischen Zeit.* Langelands Museum, Rudkøbing.
  1985 *Yngre stenalder på øerne syd for Fyn.* Langelands Museum, Rudkøbing.

Tauber, H.
  1981 $\partial^{13}C$ Evidence for Dietary Habits of Prehistoric Man in Denmark. *Nature* 292:332-333.

Thorsen, S.
  1981 "Klokkehøj" ved Bøjden. Et sydvestfunsk dyssekammer med bevaret primægrav. *Kuml* 1980: 105-146.

Troels-Smith, J.
  1953 Ertebøllekultur—Bondekultur. Resultater af de sidste 10 Aars Undersøgelser i Aamosen. *Aarbøger for Nordisk Oldkyndighed og Historie* 1953:5-62.

Vang Petersen, P.
  1984 Chronological and Regional Variation in the late Mesolithic of Eastern Denmark. *Journal of Danish Archaeology* 3:7-18.

Verhart, L.B.M.
  1990 Stone Age Bone and Antler Points as Indicators for "Social Territories" in the European Mesolithic. In *Contributions to the Mesolithic in Europe*, edited by P.M. Vermeersch and P. Van Peer, pp. 139-151. Leuven University Press, Leuven.

Wahl, J. and H. G. König
  1987 Anthropologisch-Traumologische Untersuchung der Menschlichen Skelettreste aus dem Bandkeramischen Massengrab bei Talheim, Kries Heilbronn. *Fundberichte aus Baden-Wurtemberg* 12:65-193.

Whittle, A.
  1987 Neolithic Settlement Patterns in Temperate Europe: Progress and Problems. *Journal of World Prehistory* 1:5-33.

  1988 *Problems in Neolithic Archaeology.* Cambridge University Press.

Zwelebil, M., and P. Rowley-Conwy
  1984 The Transition to Farming in Northern Europe: A Hunter-Gatherer Perspective. *Norwegian Archaeological Review* 17:104-128.

  1986 Foragers and Farmers in Atlantic Europe. In *Hunters in Transition*, edited by M. Zvelebil, pp. 67-93. Cambridge University Press.

9

The Transitions to Agriculture in Japan

Gary W. Crawford
University of Toronto

Crop production was an important part of the prehistoric economies on the Japanese archipelago from Okinawa through Hokkaidō by A.D. 1000. The transition to agriculture was not a singular process however. The standard model brings agriculture from mainland Asia as part of a diffusionary process that rapidly transformed and replaced indigenous cultures. But, this is only one aspect of the agricultural transition in Japan. There are at least four phases in the development of plant husbandry on the Japanese islands. The first phase (Transition 1) is the adoption of small scale gardening during the middle Holocene Jōmon. The next transition (Transition 2) is the disappearance of the Final Jōmon cultures with the influx of people from the mainland to southwestern Japan. The wet-rice based cultures that arose as a result of the strong influence from mainland East Asia are known collectively as the 'Yayoi.' The Tōhoku region of northeastern Japan did not accept these changes in the same way, nor as readily (Crawford and Takamiya 1990). While the Yayoi transition in the southwest involved the rapid replacement of cultures, the northeastern Japanese transformation (Transition 3) to distinctive societies together known as the 'Tōhoku Yayoi' involved the acculturation of local groups resulting in material cultures that still had marked resemblance to those of their Jōmon predecessors. The final step (Transition 4) is the replacement of the last Jōmon cultures in Japan, the Zoku Jōmon, by the Ezo-Haji or Satsumon ancestors of the Ainu.

This chapter provides an overview of the four Japanese transitions by first isolating the respective processes involved in each. Factors affecting the transitions are discussed in the context of similar transitions in eastern North America and Europe. I have been collecting primary data from Hokkaidō since 1974, so I first examine the contributions that research in Hokkaidō has made to our understanding of middle Holocene cultures, as well as late prehistoric and proto-historic cultures in northeastern Japan.

Hokkaidō Research

Salvage archaeology in Japan expanded extensively in the 1970s. The number of salvage excavations increased seven-fold to more than 7000 per year with growing land development beginning in 1969 (Tsuboi 1986:487). The number of archaeologists employed by prefectural governments correspondingly increased from about 100 to over a thousand in the same period (Kobayashi 1986; Tsuboi 1986). Minamikayabe (Figure 1), a town in southwestern Hokkaidō, began an archaeological program at the beginning of this expansive phase in Japanese archaeology. Nearly a decade and a half of excavations began at the Early Jōmon Hamanasuno site in Minamikayabe in1973. At the same time, William Hurley had been looking for an assemblage of Jōmon pottery to test analytical techniques. Yoshizaki Masakazu invited Hurley to apply his technique as a member of the Hamanasuno Project in 1974.

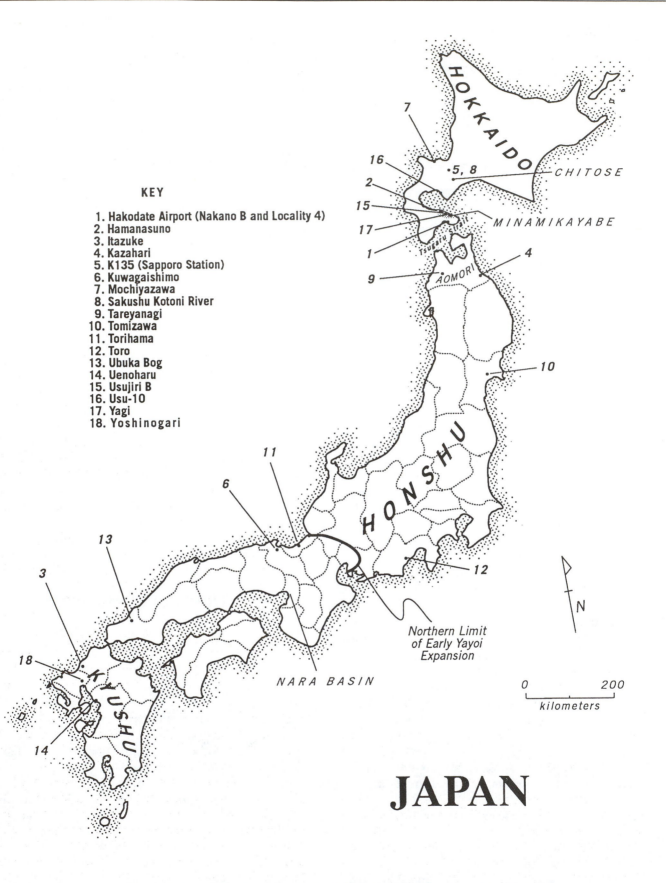

KEY

1. Hakodate Airport (Nakano B and Locality 4)
2. Hamanasuno
3. Itazuke
4. Kazahari
5. K135 (Sapporo Station)
6. Kuwagaishimo
7. Mochiyazawa
8. Sakushu Kotoni River
9. Tareyanagi
10. Tomizawa
11. Torihama
12. Toro
13. Ubuka Bog
14. Uenoharu
15. Usujiri B
16. Usu-10
17. Yagi
18. Yoshinogari

Northern Limit
of Early Yayoi
Expansion

NARA BASIN

JAPAN

Figure 1. Location of archaeological sites in Japan mentioned in the text.

Several other events in Japanese archaeology in the early 1970s made it clear that research on the analysis and interpretation of the chronological and spatial variation of ceramic motifs would need to be placed in a broader cultural context that included subsistence. Yoshinobu Kotani (1972:231-232) had announced the results of the analysis of flotation samples from the Late Jōmon Uenoharu site in western Kyūshū (Figure 1). Two cultigens were present: rice (*Oryza sativa*), and barley (*Hordeum vulgare*). At the same time a few cultigens were fortuitously recovered from a number of other Jōmon sites (Kotani 1981). Assumptions that the Jōmon was strictly a foraging adaptation had been questioned as early as the 1930s, but efforts to investigate the issue were rare. World archaeology was in the throws of a flotation revolution that was changing many assumptions about prehistoric subsistence and the relationship between people and the environment in prehistory (Watson 1976). Kotani's efforts pointed out that the situation was no different in Japan. A general evaluation of Early Jōmon subsistence at Hamanasuno was in order and subsequently undertaken in 1974 (Crawford et al. 1978; Crawford 1983; Hurley 1974). In 1976 I returned to southwestern Hokkaidō to examine Initial, Middle, and Late Jōmon sites in conjunction with further work on the Early Jōmon.

By 1977 it was clear to us that salvage archaeology in Japan, although producing extraordinary quantities of data, suffered in the same way that it does elsewhere. Substantial analysis of all classes of archaeological data in the context of specific research issues was difficult, if not impossible, to accomplish. To solve this problem, Peter Bleed and William Hurley directed excavations at the multi-component Yagi site (Figure 1) for three years under the sponsorship of Yoshizaki and Tadahisa Ogasawara. An interdisciplinary team examined settlement and community organization, subsistence, site catchment, environmental history, and Japanese research methods and epistimology (Hurley et al. 1985). Techniques included flotation, bone analysis, pollen analysis, catchment analysis, detailed attribute analysis of all classes of artifacts, use-wear analysis, and remote sensing.

Interpretations of Jōmon adaptations in Hokkaido inevitably draw upon Ainu analogies (e.g., Watanabe 1986). The Jōmon and Ainu occupied the same environmental zones, appeared to live in communities with a substantial year-round resident population, and were primarily, if not entirely, foragers who depended largely on animal food, according to these comparisons. Foraging appeared to be able to support year-round villages in Hokkaido until Japanese intercession in the late 1800s. However, the bulk of the Ainu ethnographic data are from the late nineteenth and the twentieth centuries. Although the prehistory of the Ainu was known to some extent, few archaeologists critically examined subsistence ecology, a significant aspect of any Jōmon-Ainu comparisons. We needed to test interpretations of Ainu adaptations, especially during the first millennium A.D. We had reason to be concerned about the quality of understanding of the Hokkaidō Ainu, particularly their subsistence base (Crawford and Takamiya 1990; Crawford and Yoshizaki 1987). The Ainu today perceive themselves as hunter-gatherer-fishers although they are living as contemporary Japanese in most respects. Ethnographers and historians tend to accept the Ainu self image.

Archaeological research on the early Ainu, or Ezo-Haji and Satsumon, and the Zoku Jōmon was under way in earnest by the mid-1980s. The first step involved a detailed analysis of material from the ninth century A.D. Sakushu-Kotoni River site (Figure 1) (Crawford and Yoshizaki 1987). Next, we were able to expand the first millennium A.D. data base to include the northern edge and south central region of the Ishikari Plain that surrounds Sapporo. In addition, data from eastern Aomori Prefecture are contributing to our understanding of early and pre-Ainu subsistence (D'Andrea 1992). At the same time, we have collected Early and Middle Jōmon data from the same areas that help to test our interpretations developed on the basis of the southwestern Hokkaidō studies (Crawford 1983).

Transition 1:
The Middle Holocene Record

Japan possesses a rich archaeological record of Holocene cultures. The extensive early Holocene record in Japan is particularly astounding in that, for the same period, we know little about Japan's closest geographic neighbors: China, Korea, and the Soviet Far East. Domestication processes and the evolution of agrarian societies were well under way in China by the *middle* Holocene. The archaeological record in China for this period is well known although the preceding phases are not. Although intensive food production did not develop in Korea or Japan until the first millennium A.D. (Crawford 1992), the preceding periods may provide an inferential reference for the early Holocene period in China when agriculture must have been beginning.

The earliest ceramics in Japan date to the eleventh millennium B.C. but they are not cord-marked (Esaka 1986:226). The general consensus is that the Jōmon Tradition with its cord-marked pottery begins during or towards the end of the Initial Jōmon. During the Early Jōmon, the first cultigens are arguably present but subsistence was based almost entirely on foraging (Crawford 1983; Kotani 1981).

The Yagi site has an Initial Jōmon component as does the nearby Nakano B site at Hakodate Airport (Figure 1) (Chiyo 1977; Hurley et al. 1985). Early Jōmon phases are represented in southwestern Hokkaidō at Hamanasuno as well as Yagi and Locality 4 of the Hakodate Airport site. Qualitative and quantitative differences in lithic assemblages differentiate the two periods, although similarities abound (Crawford 1983:21). Plant remains document qualitative differences between the Initial and Early Jōmon. From the late Early Jōmon to at least the beginning of the Final Jōmon, the material culture and subsistence regime is qualitatively unchanged. The Initial Jōmon people harvested nuts but the Early Jōmon inhabitants showed little interest in nuts. Weed communities were of little importance to the Initial Jōmon people in Hokkaidō, but by the end of the Early Jōmon, such communities were important collecting territories. The Early Jōmon peoples appear to have taken a keen interest in barnyard grass (*Echinochloa crusgalli*). By 2000 to 1800 B.C., the end of the Middle Jōmon at the Usujiri B site (Figure 1) a few kilometers from Hamanasuno, barnyard grass utilization increased and morphological changes in the grass fruits indicates that domestication was taking place (Crawford 1983:31-34). Other evidence for gardening includes a single carbonized grain of buckwheat (*Fagopyrum esculentum*) from an Early Jōmon house at Hamanasuno and a few grains of foxtail millet from Usujiri B. We are currently assessing the significance of the latter. In the current view, the southern Hokkaidō Jōmon seems to have been slowly developing a form of food production, but foraging remained the dominant resource procurement method for thousands of years.

Animal remains are uncommon at the sites we have been examining. The bones are usually calcined from exposure to heat, so the representativeness of the remains is more difficult to assess than it is for assemblages of unburned bone. Nevertheless, a hunting strategy has been modeled and the degree to which the animal remains from Yagi correspond to the proposed strategy has been examined (Bleed and Bleed 1981; Bleed et al. 1989). The Yagi inhabitants mainly hunted large land and sea mammals. The other remains, including some small mammals, birds and fish indicate a great deal of ecological and technological breadth, although the quantity of the small animals, estimated from the animal remains, appears lower than the model predicts (Bleed et al. 1989). The Yagi surroundings offer a rich and diverse array of resources. All the animals could have been procured within an hour or two of the community. The evidence points to butchering and processing of game within

the site limits, so there are likely no special purpose sites related to hunting (Bleed et al. 1989).

The overwhelming majority of stone tools were made of chert and coarser igneous stone whose sources are within a few kilometers of the site; much of the tool production took place on site (Bleed 1989:3). Bleed characterizes the Yagi community as one that is logistically organized but residentially immobile, as opposed to being logistically mobile (Bleed 1989:3). Some barnyard grass seeds, as well as seeds of other herbaceous annual weeds, occur in the Yagi flotation samples. This is consistent with the view that the residentially immobile Yagi population would have affected the local environment to the extent that anthropogenic communities had become significant collecting and, concomitantly, hunting territories. I would add to Bleed's comments that the Jōmon people were not passive participants in their local habitats, but were interacting with them as well; that is, their actions caused ecological changes to which the Jōmon people successfully adapted (Crawford 1983; Nishida 1981, 1983). For example, substantial disruption of local plant communities by people (anthropogenesis) resulted in early successional vegetation that was highly productive and useful. This was taking place before there is solid evidence for domestication in northern Japan.

The general patterns found in southwestern Hokkaidō are consistent with modes of subsistence, settlement, and technology found throughout Japan to the end of the Jōmon, with a few exceptions. A few small stone formations are associated with burials, and some houses or house-like structures are extremely large on many sites by the end of the Middle Jōmon. In the Late and Final Jōmon, houses are generally uniform in size and shape, but the material culture is much more elaborate than in previous periods. In addition, in northeastern Japan large stone formations, usually roughly circular, and stone-marked cemeteries abound. In the Yoichi area of Hokkaidō alone there are several hundred of these sites. Near Chitose, Hokkaidō, are several Late Jōmon earthworks. Few of these special purpose sites are associated with the Final Jōmon but the archaeological record of that period is characterized by elaborate, thin-walled pottery, pottery masks, polished stone batons, and village sites with large quantities of debris and relatively high population densities. We have yet to systematically collect plant remains from Final Jōmon sites, but work on the Late Jōmon is progressing (D'Andrea 1992).

Information from the Late Jōmon component of the Kazahari site in Aomori Prefecture, Tōhoku, (Figure 1) indicates that plant remains are not substantially different from those in the Early and Middle Jōmon of southwestern Hokkaidō. Small grass seeds are common and three cultigens, broomcorn millet (*Panicum miliaceum*), foxtail millet (*Setaria italica* ssp. *italica*), and rice, are present. A calibrated accelerator date of 787 B.C. (922 to 393 B.C. at the 67 percent confidence level and 1312 to 103 B.C. at the 95 percent confidence level) (TO-2202: 2540±240 B.P. uncalibrated) on a rice caryopsis recovered from within 5 cm of the floor of House 32, subgrid 66, confirms its Late Jōmon association (D'Andrea 1992).

The Late and Final Jōmon of southwestern Japan were not as elaborate as their contemporaries in the northeast. Populations were relatively low (Koyama 1978). Foxtail millet, rice, rice paddies, gourd (*Lagenaria siceraria*), and barley have been found at a few sites dating between 1000 B.C. and 400 B.C. (Crawford 1992; Kotani 1981). Considering the evidence in Tōhoku, broomcorn millet and rice likely were present in southwestern Japan by the second millennium B.C. Material culture with mainland connections does not occur in any abundance, however, until about 400 B.C. By 100 B.C. the Jōmon in southwestern Japan had disappeared. Shortly thereafter, two groups, the Tōhoku Yayoi (to be discussed below) and the Zoku Jōmon developed.

The Zoku Jōmon is distinctly different from its Final Jōmon predecessors. Sites such as Sapporo Station in Sapporo (Sapporo-shi Kyōiku Iinkai 1987) and Mochiyazawa in Otaru (Otaru-shi Kyōiku Iinkai 1990) are typical of Zoku Jōmon sites (Figure 1). The former is a series of short-term occupations that contain evidence of specialized resource procurement. The second site is a large cemetery with intermixed occupation debris. Few pit houses are known for the Zoku Jōmon.

Transitions 2, 3, and 4

A period of rapid transformation to substantially agrarian societies in Japan began about 400 to 300 B.C. This marks the beginning of the Yayoi Period, a wet-rice based tradition that lasted until A.D. 300. Its legacy likely still remains in the traditions unique to Japan, including the Shinto religion. The Yayoi origin has two important aspects, one being the demise of Jōmon cultures that were remarkably successful in their ability to maintain dominance in Japan for nearly 10,000 years. The other aspect is the rise of a number of distinct Yayoi groups throughout the Japanese archipelago as far north as the Tsugaru Strait that separates Hokkaidō from Honshū. A complex of major and minor crops, as well as one or two domestic animals make their appearance in Japan at this time. In southwestern Japan, agriculture appears to have been centered on wet-rice production. In the northeast, rice production developed to some extent, but in northern

Tōhoku dry land plant husbandry focussing on millets, barley, and wheat (*Triticum aestivum*) eventually emerged. This complex was carried into Hokkaidō within a few centuries. The Ainu are the descendants of these northern Tōhoku and Hokkaidō cultures (Crawford and Takamiya 1990).

Transition 2

Until recently, the origin of the Yayoi as either an indigenous transformation or the result of a massive migration from the mainland was open to question. In the first instance, archaeologists accepted that outside influence played a role, but the Yayoi was considered to be a result of acculturation of the local Jōmon and was therefore a mainly indigenous development. Hanihara (1987), Brace et al. (1989), and others now agree that the Yayoi represents a major incursion from the mainland. At first, the most pronounced continental influence was in northern Kyūshū (Kanaseki 1986). Once established in Kyūshū, local Jōmon populations appear to have coexisted with the Yayoi for a time. Little is known about the mechanisms of the replacement of the Jōmon by the Yayoi in southwestern Japan; however, the replacement was complete by 100 B.C. Complicating matters is a northwestern Kyūshū population which is interpreted to be Jōmon people who used Yayoi pottery (Brace et al. 1989:105). That is, acculturation was affecting some or all remaining Jōmon populations. Other sites such as Itazuke (Figure 1), are Jōmon occupations immediately succeeded by Yayoi occupations.

The Yayoi rapidly spread, particularly along the coasts, to the Tokai and southern Chūbu districts in central Honshū and thereafter the Yayoi advance is usually interpreted to have slowed somewhat (Akazawa 1982, 1986; Kanaseki 1986). The difference in the rate of change between the southwestern and northeastern Yayoi spread actually may not be that great. It took roughly two centuries for the Yayoi to reach central Honshū and about the same length of time for the Final Jōmon in Tōhoku to give way to the Yayoi. The rate differential may be exaggerated by those who stress the contrasting speeds of Yayoi appearance in the two areas. In other words, despite some difference in the rate of Yayoi spread, it was relatively quick in both northeastern and southwestern Japan. Without doubt the Yayoi spread is associated with a significant expansion of agriculture in Japan. Pollen diagrams consistently indicate changes such as clearance, burning, and ecological succession associated with agricultural intensification throughout southwestern Japan (Tsukada 1986; Yasuda 1978).

The disappearance of the Final Jōmon corresponds with the first two episodes of Yayoi expansion: the

Early Yayoi spread to central Honshū and the subsequent development of the Tōhoku Yayoi. The event (Transition 4) that eliminated the last of the Jōmon cultures, the Zoku Jōmon, was the movement of the Ezo-Haji (early Satsumon) ancestors of the Ainu into Hokkaidō.

The Middle Yayoi Toro site (Goto 1954) is an excellent example of the southwestern Japanese Yayoi (Figure 1). Toro combines both dwellings and ancient fields. Part of the site is waterlogged, containing well preserved wooden utensils, textiles, as well as ceramics, stone tools, and metal. Earthen embankments with wooden retaining walls form part of an irrigation network. Plant and animal remains include rice, millet, melon, gourd, peach, mammals, fresh water and marine fish, and shell fish and are evidence of a mixed economy. Although dry crops may have been grown on upland sites, a double cropping system that utilized post-harvest, drained rice paddies for a second crop was used in Japan until recently, and is still used in parts of China. Millet, barley, and wheat, for example, may have been grown as winter crops in southwestern Japan, thus permitting a full year of crop production (Crawford 1992).

Craft specialization and social differentiation are characteristics of the Yayoi. By the Middle Yayoi local areas were unifying and agricultural production was intensifying (Kanaseki 1986:319). Imported mirrors and iron weapons became more abundant as well (Kanaseki 1986:331). Long distance movement of goods and/or people within Japan is evidenced by, for example, Okinawan shell comprising nearly all the jewelry uncovered at the Usu-10 site in southwestern Hokkaidō (Figure 1). Inter-group hostilities in southwestern Japan are evidenced but disappear by the end of the Yayoi (Ikawa-Smith 1985:394).

The Tōhoku Yayoi was in many ways Jōmon in character (Crawford and Takamiya 1990; Ito 1986), but it is distinguished from the preceding Final Jōmon by intensive food production. Rice paddies have been found at the first century A.D. Tareyanagi and Tomizawa sites (Figure 1) (Ito 1986). The Yayoi occupation at the Kazahari site contains rice, broomcorn millet, and foxtail millet (D'Andrea 1992). The northern Yayoi was probably never as dependent on rice as its southwestern counterpart. Barley, wheat, and millets were important and eventually became the dominant crops in prehistoric and protohistoric northeastern Japan (Crawford and Takamiya 1990).

The intensive food production of the Yayoi period eventually enabled the support of a full-time, centralized government. The first indication of the end of the Yayoi is the appearance of large tombs (*kofun*), the largest of which are keyhole shaped. The Yamato state, marking the beginning of the Japanese nation, arose in

the Nara Basin during the Kofun Period (A.D. 300–700), named for its characteristic earthen tombs. The Kofun Period ends a century or so after the emergence of recognizable Ainu ancestors in the north.

The Yamato state emerged during the Kofun Period by the early 6th century A.D. One of the outcomes of the Yamato state formation was the institutionalization of occupational groups (*be*) (Barnes 1987:86). The administrative organization of this state was a structure that provided direct access to the commoner, breaking up the locally autonomous cells characteristic of the Yayoi (Barnes 1987). Mechanisms such as the integration of occupational groups assured continued dependence of the various regions on the state (Barnes 1987:46). Farmers were one such specialized occupation group. The Kofun Period is associated with a new era of interaction with the mainland and exotic items appear in Japan at this time. One important introduction was the horse and another was Buddhism (Ikawa-Smith 1985). By the end of the Kofun Period, materials from as far away as western Asia were moving into Japan. Domesticated plants such as safflower (*Carthamus tinctorius*) from the Middle East were introduced at this time (Crawford and Takamiya 1990).

Transition 3

Northern Honshū came under some undefined influence of the Yamato state by the seventh century (Aikens and Higuchi 1982; Crawford and Takamiya 1990). Shortly after the Yamato state emerged in southwestern Japan, northern Yayoi pottery was supplanted by Haji ware typical of the Kofun Period. The ceramics were all locally manufactured, however. Numerous keyhole-shaped tombs like those in southwestern Japan are found as far north as Tōhoku; circular tombs are found in central Hokkaidō. House styles were similar from Hokkaidō to the Kantō district at this time. The Yamato state may have been trying to incorporate the autonomous northern cultures into its fold in much the same manner as Barnes (1987) describes for southwestern Japan. Ikawa-Smith has suggested that ideology in the form of a *kofun* cult was an important aspect of the spread of Yamato influence and that the appearance of *kofun* throughout Japan need not indicate local ties to a central Yamato authority (1985:395). Certainly the Yamato state did not incorporate Tōhoku as it did southwestern Japan. In the first half of the second millennium A.D. authorities representing the Japanese government were in place in Tōhoku, but the region was largely foreign territory to the Japanese. The local inhabitants, called the *emishi* or *Ezo* (Aston 1896:XXVI,262), are generally accepted to be the ancestors of the Ainu. Following the precedent set by Yoshizaki (1984), I refer to the archaeological cultures of the period from about A.D. 600 to A.D. 900 as the Ezo-Haji, a name reflecting the likely cultural identity of the people as well as the ceramic complex in use at the time. Others refer to this phase as early Satsumon (Aikens and Higuchi 1982). Not until the middle of the second millennium A.D. did the Japanese state come to include Tōhoku and Hokkaido (for a more detailed discussion see Crawford and Takamiya 1990).

Transition 4

Transition 4 took place some time shortly after A.D. 500–600 (Crawford and Takamiya 1990). By A.D. 700–800 only dry crop agriculture was being carried out by the Ezo-Haji throughout the Ishikari Plain in central Hokkaidō, as well as in southern Hokkaidō. This pattern correlates well with records of an exodus of native peoples from Tōhoku (Takakura 1960). So far, there is no evidence of Ezo-Haji artifacts at Zoku Jōmon sites or the reverse. Hanihara (1990a) prefers a unilineal scheme that derives the Ainu from Zoku-Jōmon ancestors because of the physical similarity between the Jōmon and Ainu populations in the north. Hanihara's model assumes that the Ainu and their direct ancestors were hunter-gatherers. The model of Ainu origins deriving them from Tōhoku rather than the Hokkaidō Zoku Jōmon is consistent with the human osteological data as well as the archaeological data on agriculture and technology of the Ezo-Haji.

Ezo sites are all small hamlets or villages. Ceramics are generally pots with flaring rims and bowls (Figure 2). There is very little geographic variation in pottery styles at this time. Variation through time is recognized, particularly in the more complex trailing/incising that becomes more common in later periods around the rims of pots. In contrast, the crops vary from region to region. On the Northern Ishikari Plain, the west coast of Hokkaidō and in southern Hokkaidō, a wide variety of crops are found in the plant remains. In the southern Ishikari Plain, almost all the cultigen seeds we have recovered so far are foxtail and broom-corn millet (Crawford and Takamiya 1990:905; Crawford 1991). At the moment we have no explanation for this pattern.

Conditions and Causal Factors

A number of proximate causes have been suggested to have given rise to food production in Japan in particular and East Asia in general. These include deterministic environmental changes, demographic influences, and/or social forces. Little progress has been made in exploring complex causality in the way

that Flannery (1986) has done for Mexico, however. This is partly because the concept of diffusion dominates thinking about agricultural origins in Japan; but the factors giving rise to agriculture in primary areas, as opposed to those influencing developments in secondary regions, should be carefully considered.

Eurasia, North America, and Japan are among the best documented of the regions in which agricultural origins or intensification is primarily a result of diffusionary processes (e.g., Smith 1990; Watson 1989; Zvelebil 1986). I include *intensification* because in North America and Japan, in particular middle Holocene domestication and plant husbandry appear long before more intensive forms of food production develop. Intensive food production spreads in eastern North America during the latter half of the first millennium A.D., after the introduction of corn (*Zea mays*). (In Illinois and Ohio intensification of non-corn food production may well precede the introduction of corn.) In Japan, intensified food production is first associated with rice paddy agriculture in the first millennium B.C. Among the proposed factors influencing the rate of the transition from foraging to farming and the ultimate replacement of foraging by farming in Eurasia are: climate; colonization and a related population imbalance between foragers and farmers; social structure and the emergence of ranking; and the existence of a frontier along which disruption of foraging patterns occurred (Gebauer and Price 1990; Zvelebil 1986:183). In eastern North America, familiarity with plant husbandry may be added to the list, although this does not appear to be a factor on the northern peripheries such as in southern Ontario, Canada. These factors may have been involved in the transition to food production in Japan as well.

The Japanese situation, although sharing a number of characteristics with prehistoric eastern North American and European cultures, is unique in several ways. Material culture, including pottery, was elaborate and sedentism was much more widespread in contrast to the eastern North American Archaic and the European Mesolithic. Animal husbandry was important in Neolithic Europe, but not in Japan or eastern North America. Indigenous plant husbandry was apparently a component of many early eastern North American and Japanese economies, but not of European Mesolithic economies.

The natural environment is often viewed as a determining factor in regard to Jōmon culture. This can be a productive line of enquiry, as in the examination of the convergence of the Kantō Jōmon and the central California Windmiller and later cultures (Aikens and Dumond 1986). In other cases, however, problems arise. Early views of an indigenous Jōmon origin of agriculture in Japan were based on contradictory

Figure 2. Ezo-Haji pot from the Sakushu-Kotoni River site, Hokkaidō. Height: approximately 28 cm.

notions of the richness and the scarcity of resources in the Japanese environment. The comparatively high population density of the Chūbu district Middle Jōmon (central-western Japan), despite the rich resources of the area, could not be attributed entirely to a foraging culture (Fujimori 1963). The elaborate stone and ceramic technology seemed out of place in a foraging context, according to Tsuboi (1964) and Fujimori (1963), and may well have been the result of an agricultural economy.

In contrast, Nakao (1966) and Ueyama (1969) proposed that the sparsely populated western Japanese Jōmon was involved with a form of incipient food production. The low density populations resulted from the scarce resources of the broad-leaf evergreen forest zone. To them, the swidden systems of mainland Asia in the same forest zone implied that the western Jōmon either carried out "semi-cultivation" or swidden in order to solve the resource scarcity problem (Nakao 1966; Sasaki 1971; Ueyama 1969). Scarcity, therefore, provided a *need* for agriculture. In part, these views might best be characterized as wishful thinking.

In recent years the concept of affluent foragers has pointed to the elaboration of culture in resource rich environments without the presence of food production (e.g. Koyama and Thomas 1981; Price and Petersen 1987). Aikens (1981) and Hayden (1990 and this volume) go so far as to propose that food production arose because a social elite among foragers demanded increased production to serve their needs. These demands would lead to food production when the system reached its carrying capacity. In the common wisdom, we think of food production as necessary for elite classes to become established. Later in this chapter I examine this theme of social elaboration in more detail, but in terms of resource abundance or scarcity as a factor in agricultural origins, the seasonal cycle of resources and their annual unpredictability is at issue, not their absolute or average abundance. The scarcity or richness issue is not adequately resolved unless we can measure resources in the context of population densities and technology.

Environmental change during the Holocene likely had an impact on Jōmon subsistence. In the southwest, Yasuda (1978:254) interprets pollen evidence to mean that an *Artemisia* steppe developed after the Pleistocene boreal forests disappeared about 10,000 B.P. The sparsely represented Incipient Jōmon may have developed in this environment. Broad-leaf forests replaced the steppe by 8500 B.P., first with deciduous then with evergreen broad-leaf species (Yasuda 1978). The development of the Initial Jōmon by 8500–8000 B.P., presumably from an Incipient Jōmon ancestor, coincides roughly with the establishment of broad-leaf

forests throughout Japan. Chard's (1974) co-traditions were in place; the southwestern Early Jōmon was well developed by then, with characteristics differentiating it from the northeastern Early Jōmon.

In the northeast, where no Incipient Jōmon is known, the boreal forest was replaced by oak (*Quercus*) and beech (*Fagus*) forest between 9000 and 8000 B.P. (Tsukada 1986:27-28). Tsukada (1986:28) proposes that a clearer understanding of Jōmon subsistence will derive from understanding beech distribution dynamics through time because beech nuts were an important Jōmon food. Tsukada is mistaken. We have no evidence that beech nuts were a food, let alone an important resource, during the Jōmon period. The evidence for acorns is not much better. Much as in eastern North America, people were not concerned with beech nuts, a resource that is relatively inefficient to harvest. Understanding the long-lasting subsistence regime in Hokkaidō that began in the Early Jōmon is much more complex than examining distributions of beech trees. The establishment of deciduous broad leaf forests in southern Hokkaidō is followed by decreased nut use and increased anthropogenic plant use by 6000 to 5000 B.P. (Crawford 1983).

Ho (1977) proposed that the steppe-forest ecotone played a role in the origins of agriculture in China, although he did not elaborate. I have cautioned elsewhere (Crawford 1992) that borrowing this model to explain developments in Japan would be futile. Ho's evidence for a steppe or a steppe-forest origin of agriculture in China is inadequate. The evidence for an early Holocene steppe in Japan is not correlated with a substantial move to food production. The evidence for subsistence change, which may have had food production as a component, is later than 10,000–8500 B.P. on the whole, and long after the broad-leaf forests were established throughout their modern range in Japan.

Human induced changes in local ecology, or anthropogenesis, take two forms pertinent to agriculture. The first involves the disruption of habitats due to agricultural activities. The second involves the non-agricultural creation of open, disturbed habitats that are colonized by weeds as a proposed initial step in the domestication process. In the first case, by tracing the history of changes known to be the result of cultivation, the development and spread of agriculture can be discerned (e.g., Dimbleby 1978). In the second case, some weeds may be potential cultigens and are thought to be ancestral to domesticated plants (Anderson 1971; Harlan and de Wet 1965). Domesticated plants may not all have weeds as their ancestors however. Hillman and Davies (1990) observed no wild barley or wheat colonizing humanly disturbed areas in western Asia and believe that agricultural activities preceded domestication there. In any case, evidence of environmental disruption may provide important clues to the domestication process and the origins of agriculture.

Certain evidence in the vegetation history of Japan is related to agricultural activities. Tsukada (1986) links the development of pine (*Pinus*) woodlands and forest fires evidenced in pollen cores to agriculture in Japan. The pine forest induced by humans as a secondary succession stage due to clearance for agriculture is probably a relatively late phenomenon, developing "a few thousand years ago" (Tsukada 1986:46). Frequent forest fires caused by people, but not related to agriculture are evidenced as early as 8500-7000 B.P. in southwestern Japan (Tsukada 1986:41). However, at Ubuka Bog, shifting agriculture appears about 7700 B.P. and pollen from buckwheat, an exotic cultigen, is common in levels dated to 6600 B.P. (Tsukada et al. 1986; Tsukada 1986).

The data from southwestern Hokkaidō are similar to Tsukada's. Rather than examine regionally sensitive pollen evidence, I have attempted to isolate the plant remains at the sites that were remnants of plant parts harvested by people (Crawford 1983). From the Early Jōmon to the end of the Middle Jōmon, the latest period in the Minamikayabe study, no major plants were added to, or removed from, the Jōmon diet (Crawford 1983). The evidence does indicate an increasing dependence on annual and perennial weedy taxa such as knotweeds (*Polygonum*), grasses (Gramineae), sumac (*Rhus*), and other forest edge and open ground plants. Site specific pollen records imply large-scale local disruptions, but these disturbances do not show up on regional pollen diagrams (Davis 1979). At Yagi, for example, non-arboreal plants had greater input to the pollen record than arboreal pollen did (Davis 1979). In my view, disturbed habitats, whether gardens or not, were important to the Jōmon. Through time they became increasingly dependent on resources from such habitats.

Nishida (1983) provides a slightly different twist to the discussion of anthropogenesis. In his opinion, the Jōmon developed a symbiotic relationship with chestnut (*Castanea*) and walnut (*Juglans*) trees as well as a host of other sun-loving plants (Nishida 1983:315). Nut trees were by far the most important of these plants, however. The nature of the deposition of the plant remains at the waterlogged sites Nishida has studied (Torihama and Kuwagaishimo) makes it almost impossible to ascertain which remains are food and which are not. Nevertheless, Nishida and I agree on the importance of anthropogenic communities to the Jōmon people, although we do not agree on the specific plants that were used.

The *success* of indigenous cultures is a common theme helping to explain a relatively slow rate of the transition to, or intensification of, food production. Underlying themes include the degree to which foraging cultures were disrupted or disruptable by new ideas, people, and technology. Gregg (1988) models a symbiotic relationship between foragers and farmers that involves an exchange of food resources. Foragers successfully maintained their existence by cooperating with farmers, according to this model. Unfortunately, Gregg is unable to find good archaeological data to test her model. Keely (this volume) interprets the archaeological record for Linearbandkeramik to mean that the two groups of farmers and foragers in fact repelled each other and never assimilated. Keely sees no evidence of exchange or any other kind of cooperation. Gebauer and Price (1990) also do not support the interaction model. The cooperative model usually involves the exchange of crops and domestic animals for service or wild resources. They report, however, that crops and domesticated animals were the last items to be borrowed by Mesolithic foragers in Denmark (Gebauer and Price 1990:260).

European archaeologists admit to some frustration with the archaeological data from the period of the Mesolithic demise. Radiocarbon dates are confusing and little organic material has been recovered from sites of this period (Gregg 1988; Keely this volume). Japan, on the other hand, offers a well established chronology, good human osteological data, and a plethora of sites rich in organic debris.

Whatever the natural environmental circumstances and the success of the Jōmon, little doubt remains that the rapid spread of the Yayoi and the end of the Jōmon in southwestern Japan was a result of the reproductive and technological success of these newcomers to Japan. The Yayoi in the southwest are physically much like mainland northeast Asian peoples (Brace et al. 1990; Hanihara 1987, 1990a and b). Hanihara (1990a; 1990b) argues that the southwestern Japanese and mainland Asians are not entirely alike; the variation from mainland-like Japanese to Ainu-like Japanese follows a cline from southwest to northeast in Japan. The Kyūshū Yayoi is particularly northeast Asian in appearance. One major Yayoi site, Yoshinogari (Figure 1), excavated recently was planned by the migrants (Hanihara 1990a). If little acculturation took place between the Yayoi and Jōmon in southwestern Japan, Akazawa's (1982; 1986) argument that the southwestern Jōmon was predisposed to accept rice production because of their familiarity with gardening may be moot. No matter what the Jōmon adaptation in southwestern Japan, replacement and acculturation appear to be the main process of agricultural development.

Relative population densities and the nature of local economies and sociocultural systems are significant issues in explaining the initial halt of the early Yayoi spread at the Tokai boundary, followed by the development of the Tōhoku Yayoi and eventually an agricultural system based not on rice, but on wheat, barley, and millets among other crops (Crawford and Takamiya 1990). The northern Yayoi and subsequent cultures retained much of their Jōmon physical characteristics. The osteological pattern of the Ainu, who descended from these cultures in northern Tōhoku (Crawford and Takamiya 1990), differs little from the Jōmon pattern (Brace et al. 1990). The mechanisms for the development of the northern Yayoi have not been explored, but are the subject of current research (D'Andrea 1992).

Material evidence of elaborate public works and ritual are characteristic of the Late and Final Jōmon of Tōhoku and Hokkaidō, as I have already mentioned. This sort of archaeological evidence is usually interpreted to be evidence of social ranking and some degree of public economy. Some have argued that it is the very existence of such structures that is instrumental in the adoption of food production (for a summary, see Gebauer and Price [1990:264]).

Aikens (1981), in a comparison of the prehistory of Japan and eastern North America, concludes that in both areas the change was from broad-spectrum forest economies to agriculture. He traces the evidence for increasing societal complexity and finds that sociopolitical structure that evolved in part due to rich woodland resources ultimately created demand for greater resources and that this demand was satisfied by agriculture. Aikens proposed that societal organization was the crucial factor in the rise of agriculture in both areas (1981:271). Whatever the merit of the original argument, the current data are evidence for the opposite; cultigens precede the development of social complexity. In eastern North America a grain crop, sumpweed (*Iva annua*), was extensively harvested during the Middle Archaic and was domesticated by the Late Archaic in the Illinois Valley (Asch, and Asch 1978; Asch 1985). Several other cultigens are present by the Late Archaic and Early Woodland including chenopod (*Chenopodium berlandiari* ssp. *jonesianum*), cucurbit (*Cucurbita pepo*), sunflower (*Helianthus annuus*), bottle gourd (*Lagenaria siceraria*), and possibly maygrass (*Phalaris caroliniana*) (Fritz 1990; Smith 1990; Watson 1989). Thus, domesticated grain plants, as well as other crops, appear in the archaeological record of eastern North America before there is substantial evidence for changes in social organization related to the establishment of conspicuous consumption.

Unfortunately, the situation in northern Japan is not quite as well researched as it is in eastern North America. I have argued that the Early and Middle Jōmon plant food subsistence has a number of parallels with that of the Early to Late Woodland of North America (Crawford 1983). Domestication of barnyard grass was under way by 2000 B.C. (Crawford 1983) and three cultigens have been identified from a Late Jōmon site in northern Tōhoku (D'Andrea 1992). We have no way of knowing, however, whether they were grown there or imported to Tōhoku. Once the Tōhoku Yayoi was established, the Final Jōmon disappeared from Hokkaidō and was followed by the Zoku Jōmon which lacks any of the extensive material culture and site elaboration of its predecessors. All available evidence indicates a *collapse* of the public economy in Hokkaidō with the introduction of intensive food production, as witnessed by the Zoku Jōmon successors to the Final Jōmon. Cultigens are being recovered from Zoku Jōmon sites, but we do not know whether they were grown in Hokkaidō (Crawford and Takamiya 1990; D'Andrea 1992).

The Tōhoku Yayoi socioeconomic organization is not well known, but because chiefdoms were present in southwestern Japan, they were likely in Tōhoku as well. If this is the case, the Final Jōmon in Tōhoku, with its evidence for social ranking and public economy, was succeeded by local Yayoi societies who incorporated social ranking and public economy to a greater degree than their predecessors. However, rice and two millets were present in Tōhoku at least by the end of the Late Jōmon and barnyard millet and buckwheat and perhaps foxtail millet were present as early as the Early to Middle Jōmon in neighboring southwestern Hokkaidō. It is difficult to reconcile the socioeconomic elaboration model with the available data in northeastern Japan. That is, it seems likely that grain foods were present before the elaboration seen during the Final Jōmon and grains were certainly available to the late Late Jōmon cultures.

Concluding Remarks

Japanese prehistory offers a number of opportunities to examine the transitions to agriculture in a relatively unique cultural and ecological setting. Two regions with similar processes of agricultural development are Europe and eastern North America. Yet differences among the three areas are significant. To clarify the processes in Japan, I have discribed four phases, Transitions 1 through 4, of agricultural development there. The first transition, which is the least well documented of the four, is the largely indigenous use of gardens during the middle Holocene Jōmon. Many of the cultigens such as buckwheat and bottle gourd reported from middle Holocene Jōmon sites are exotic to Japan, yet at least one, barnyard grass/millet, is not. Much as in eastern North America, little if any other material evidence of external contact has been found besides the cultigens. The evidence of local and regional ecological changes that have been attributed to human influences is also suggestive. These changes begin before the first cultigens appear in the archaeological record.

The second transition that resulted in the Yayoi culture is clearly a result of influence external to Japan, including human migration and the reproductive success of the Yayoi peoples. The third transition brought agriculture to northeastern Japan, but wholesale replacement of indigenous cultures does not appear to have occurred there. The change appears to have created local Yayoi groups who maintained much of the identity of their predecessors. The mechanisms for this change are not obvious. How the change took place at the rate that it did and at the time that it did are areas for further research. Models of social interaction and disruption may well apply.

The final northward expansion of an agricultural way of life took place in the latter half of the first millennium A.D. after a distinctive northern agricultural complex developed. Evidence indicates a replacement of Zoku Jōmon populations in Hokkaido by human populations biologically similar to the Jōmon but culturally quite different. The newcomers, the Ezo-Haji or early Satsumon, were ancestors of the Ainu. The precise mechanisms of the change from Zoku Jōmon to early Satsumon are not known. Although the reproductive success and technological advantages of the Satsumon are likely important reasons for their relative success, it is still quite possible that acculturation of some Zoku Jōmon peoples took place. Nevertheless, by A.D. 1000 two distinct systems of agriculture developed in Japan. One was based on wet rice production and the other on millets, barley, wheat, and a host of other dry field crops.

Acknowledgements

My Hokkaidō research has been sponsored by the Social Sciences and Humanities Research Council of Canada (Grant No. 410-86-0769 and 410-89-0786); a bilateral exchange grant from SSHRC and the Nihon Gakujitsu Shinkōkai (Grant No. 473-86-0007); and two grants from Earthwatch. A version of this paper was presented at the 56th Annual Meeting of the Society for American Archaeology in New Orleans, April 1991. The University of Toronto, Erindale College, provided a travel grant to assist in my attendence at the conference.

The continuing support of many colleagues in Japan has made our research possible. In particular, Masakazu Yoshizaki has been more generous of his time and resources than anyone has the right to expect. He has acted as mentor and facilitator to all the western archaeologists who have worked on the various Hokkaidō projects. Three graduate students have had a significant role in the success of the last five years of field research. They are A.C. D'Andrea, Keven Leonard, and Hiroto Takamiya. Of course, there are many more people and organizations to credit, including Yasuyo Tsubakisaka, Hokkaidō University, the Hokkaidō Prefectural Board of Education, the Prefectural Salvage Archaeology Centers, and several municipal and town Boards of Education (Minamikayabe, Hachinohe, Sapporo, Chitose, Eniwa, Otaru and Yoichi). I also wish to thank the editors of this book, Anne Birgitte Gebauer and T. Douglas Price, for their constructive help, discussions, and invitation to contribute this paper.

References Cited

Aikens, C. Melvin
1981   The Last 10,000 Years in Japan and Eastern North America: Parallels in Environment, Economic Adaptation, Growth of Societal Complexity, and the Adoption of Agriculture. In *Affluent Foragers*, edited by S. Koyama and D.H. Thomas, pp. 261–273. Senri Ethnological Studies 9, National Museum of Ethnology, Osaka.
Aikens, C. Melvin, and Don E. Dumond
1986   Convergence and Common Heritage: Some Parallels in the Archaeology of Japan and Western North America. In *Windows on the Japanese Past: Studies in Archaeology and Prehistory*, edited by Richard Pearson, pp. 163–178. Center for Japanese Studies, University of Michigan, Ann Arbor.
Aikens, C. Melvin, and Takayasu Higuchi
1982   *The Prehistory of Japan.* Academic Press, N.Y.
Akazawa, Takeru
1982   Cultural Change in Prehistoric Japan: Receptivity to Rice Agriculture in the Japanese Archipelago. In *Advances in World Archaeology Vol. 1,* edited by Fred Wendorf and Angela E. Close, pp. 151–211. Academic Press, N.Y.

1986 Hunter-Gatherer Adaptations and the Transition to Food Production in Japan. In *Hunters in Transition*, edited by Marek Zvelebil, pp. 151–165. Cambridge University Press, London.
Anderson, Edgar
1971 *Plants, Man, and Life.* University of California Press, Berkeley.
Asch, Nancy B. and David L. Asch
1978 The Economic Potential of *Iva annua* and Its Prehistoric Importance in the Lower Illinois Valley. In *The Nature and Status of Ethnobotany*, edited by Richard I. Ford, pp. 301–342. Anthropological Papers No.67, Museum of Anthropology, University of Michigan, Ann Arbor.
1985 Prehistoric Plant Cultivation in West-Central Illinois. In *Prehistoric Food Production in America*, edited by Richard I. Ford, pp. 149–203. Anthropological Papers No. 75, Museum of Anthropology, University of Michigan, Ann Arbor.
Aston, W.G.
1896 Nihongi, Chronicles of Japan from the Earliest Times to A.D. 697. Charles E. Tuttle, Tōkyo (1972).
Barnes, Gina L.
1987 The Role of the *be* in the Formation of the Yamato State. In *Specialization, Exchange, and Complex Societies*, edited by Elizabeth M. Brumfiel and Timothy K. Earle, pp. 86–101. Cambridge University Press, London.
Bleed, Ann Salomon, and Peter Bleed
1981 Animal Resources of the Yagi Community: a Theoretical Construction of Early Jōmon Hunting Patterns. Technical Report 81–06, Division of Archeological Research, University of Nebraska-Lincoln.
Bleed, Peter
1989 Ready for Anything: Technological Adaptations to Ecological Diversity at Yagi, an Early Jōmon Community in Southwestern Hokkaidō, Japan. Paper Presented at the Circum-Pacific Prehistory Conference, Seattle, Washington.
Bleed, Peter, Carl Falk, Ann Bleed, and Akira Matsui
1989 Between the Mountains and the Sea: Optimal Hunting Patterns and Faunal Remains at Yagi, an Early Jōmon Community in Southwestern Hokkaidō. *Artic Anthropology* 26: 107–126.
Brace, C. L., M. L. Brace, and W. R. Leonard
1989 Reflections on the Face of Japan: a Multivariate Craniofacial and Odontometric Perspective. *American Journal of Physical Anthropology* 78:93–113.
Chard, Chester S.
1974 *Northeast Asia in Prehistory.* University of Wisconsin Press, Madison.
Chiyo, Hajime (editor)
1977 Hakodate Kūkō Dai-4-Chiten to Nakano Iseki (Hakodate Airport Locality 4 and the Nakano Site). Hakodate Kyōiku Iinkai, Hakodate-shi.
Crawford, Gary W.
1983 *Paleoethnobotany of the Kameda Peninsula Jōmon.* Anthropological Papers No. 73. Museum of Anthropology, University of Michigan, Ann Arbor.
1991 The North Asian Plant Husbandry Project. Unpublished SSHRC Grant Report, Ms. in possession of the author.
1992 Prehistoric Plant Domestication in East Asia: the Japanese Perspective. In *The Origins of Plant Domestication in World Perspective*, edited by Patty Jo Watson and C. Wesley Cowan. Smithsonian Institution Press, Washington D.C., in press.
Crawford, Gary W., William Hurley, and Masakazu Yoshizaki
1978 Implications of Plant Remains from the Early Jōmon Hamanasuno Site, Hokkaidō. *Asian Perspectives* XIX:145–155.
Crawford, Gary W., and Hiroto Takamiya
1990 The Origins and Implications of Late Prehistoric Plant Husbandry in Northern Japan. *Antiquity* 64:889–911.
Crawford, Gary W., and M. Yoshizaki
1987 Ainu Ancestors and Early Asian Agriculture. *Journal of Archaeological Science* 14:201–213.

D'Andrea, A.C.
  1992 Paleoethnobotany of Later Jōmon and Early Yayoi Cultures in Northeastern Japan: Northeastern Aomori and Southwestern Hokkaidō. Unpublished PhD. dissertation, University of Toronto, Toronto.

Davis, A.M.
  1979 Clearance and Agriculture and its Influence on Regional and Local Pollen Rains in the Jōmon and Yayoi Periods with Particular Reference to Northern Honshū and Southwestern Hokkaidō. Paper presented at the 44th Annual Meeting of the Society for American Archaeology, Vancouver.

Dimbleby, Geoffrey
  1978 *Plants and Archaeology.* Humanities Press, N.J.

Esaka, Teruya
  1986 The Origins and Characteristics of Jōmon Ceramic Culture: A Brief Introduction. In *Windows on the Japanese Past: Studies in Archaeology and Prehistory,* edited by Richard Pearson, pp. 223–228. Center for Japanese Studies, University of Michigan, Ann Arbor.

Flannery, Kent
  1986 *Guila Naquitz.* Academic Press, New York.

Fritz, Gayle J.
  1990 Multiple Pathways to Farming in Precontact Eastern North America. *Journal of World Prehistory* 4:387–435.

Fujimori, Eiichi
  1963 *Jōmon Jidai Nōkōron to Sono Tenkai* (The Argument for Jōmon Agriculture and Its Development). *Kokogaku Kenkyu* 10:322–339.

Gebauer, Anne Birgitte, and T. Douglas Price
  1990 The End of the Mesolithic in Eastern Denmark: a Preliminary Report on the Saltbæk Vig Project. In *Contributions to the Mesolithic in Europe,* edited by Pierre M. Vermeersch and Philip Van Peer, pp. 259–280. Leuven University Press.

Goto, S.
  1954 *Toro Iseki (The Toro Site).* Nihon Kōkogaku Kyōkai-ken and Mainichi Shinbunsha, Tōkyō.

Gregg, Susan
  1988 *Foragers and Farmers.* The University of Chicago Press, Chicago.

Hanihara, Kazuro
  1987 Estimation of the Number of Early Migrants to Japan: a Simulative Study. *Journal of the Anthropological Society of Nippon* 95:391–403.
  1990a Dual Structure Model for the Population History of the Japanese. *Japan Review* 2:1–33.
  1990b Emishi, Ezo, and Ainu: an Anthropological Perspective. *Japan Review* 1:35–48.

Harlan, J. R., and J.M.J. de Wet
  1965 Some Thoughts About Weeds. *Economic Botany* 19:16–24.

Hayden, Brian
  1990 Nimrods, Piscators, Pluckers and Planters: the Emergence of Food Production. *Journal of Anthropological Archaeology* 9:31–69.

Hillman, Gordon, and M. Stuart Davies
  1990 Measured Domestication Rates in Wild Wheats and Barley Under Primitive Cultivation, and Their Archaeological Implications. *Journal of World Prehistory* 4:157–222.

Ho, Ping-Ti
  1977 The Indigenous Origins of Chinese Agriculture. In *Origins of Chinese Agriculture,* edited by C.A. Reed, pp. 413-484. Mouton Publishers, Chicago.

Hurley, William M.
  1974 The Hamanasuno Project. *Arctic Anthropology* XI (Supplement):171–176.

Hurley, William M., Masakazu Yoshizaki, Peter Bleed, and John Weymouth
  1985 Early Jōmon Site at Yagi, Hokkaidō, Japan. *National Geographic Society Research Reports* 19:365–381.

Ikawa-Smith, Fumiko
  1985 Political Evolution in Late Prehistoric Japan. In *Recent Advances in Indo-Pacific Prehistory,* edited by V.N. Misra and Peter Bellwood, pp. 391–398. Oxford and IBH Publishing Co., New Delhi.

Ito, Nobuo
  1986 Tōhoku Chiho ni Okeru Inasaku Nōkōno Seiritsu (The Emergence of Rice Agriculture in Tōhoku). In *Nihonshi no Reimei,* edited by N. Egami, pp. 335–365. Rokko Shuppan, Tokyo.

Kanaseki, Hiroshi
  1986 The Evidence for Social Change Between the Early and Middle Yayoi. In *Windows on the Japanese Past: Studies in Archaeology and Prehistory,* edited by Richard Pearson, pp. 317–333. Center for Japanese Studies, University of Michigan, Ann Arbor.

Kobayashi, Tatsuo
  1986 Trends in Administrative Salvage Archaeology. In *Windows on the Japanese Past: Studies in Archaeology and Prehistory,* edited by Richard Pearson, pp. 491–496. Center for Japanese Studies, University of Michigan, Ann Arbor.

Kotani, Yoshinobu
  1972 *Economic Bases During the Later Jōmon Period in Kyūshū, Japan: A Reconsideration.* Ph.D. dissertation, University of Wisconsin. University Microfilms, Ann Arbor.
  1981 Evidence of Plant Cultivation in Jōmon Japan: Some Implications. In *Affluent Foragers,* edited by S. Koyama and D.H. Thomas, pp. 201–212. Senri Ethnological Studies 9. Senri Museum of Ethnology, Osaka.

Koyama, Shuzo
  1978 *Jōmon Subsistence and Population.* Senri Ethnological Studies, Miscellanea 1. National Museum of Ethnology, Osaka.

Koyama, Shuzo and David H. Thomas (editors)
  1981 *Affluent Foragers.* Senri Ethnological Studies No. 9. National Museum of Ethnology, Senri, Osaka.

Nakao, Sasuke
  1966 Saibai Shokubutsu to Noko no Kigen (Cultigens and the Origin of Agriculture). Iwanami Shinsho, Tokyo.

Nishida, Masaki
  1981 Jōmon-Jidai no Ningen-Shokubutsu Kankei Shokuryo Seisan no Shutsugen Katei (Human-Plant Relationships in the Jōmon Period and the Emergence of Food Production). *Bulletin of the National Museum of Ethnology* 6:234–255.

Nishida, Masaki
  1983 The Emergence of Food Production in Neolithic Japan. *Journal of Anthropological Archaeology* 2:305–322.

Otaru-shi Kyōiku Iinkai (editor)
  1990 *Ranshima Mochiyazawa Iseki (The Ranshima Mochiyazawa Site).* Otaru-shi Maizo Bunkazai Chōsa Hokokusho, 2. Otaru-shi Kyōiku Iinkai, Otaru-shi.

Price, T. Douglas, and Erik Brinch Petersen
  1987 A Mesolithic Camp in Denmark. *Scientific American* 256:113–121.

Sasaki, Komei
  1971 *Inasaku Izen (Before Rice Agriculture).* Nihon Hoso Shuppan Kyokai (NHK Books) No. 147, Tōkyō.

Sapporo-shi Kyōiku Iinkai
  1987 *K135 Iseki (The K135 Site).* Sapporo-shi Bunkazai Chōsa Hōkokusho XXX. Sapporo-shi Kyōiku Iinkai, Sapporo-shi.

Smith, Bruce
  1990 Origins of Agriculture in Eastern North America. *Science* 246:1566–1571.

Takakura, Shinichiro
  1960 The Ainu of Northern Japan. *Transactions of the American Philisophical Society* 50(4). Philadelphia.

Tsuboi, Kiyotari
  1986 Problems Concerning the Preservation of Archaeological Sites in Japan. In *Windows on the Japanese Past: Studies in Archaeology and Prehistory,* edited by Richard Pearson, pp. 481–490. Center for Japanese Studies, University of Michigan, Ann Arbor.

Tsukada, M.
   1986   Vegetation in Prehistoric Japan. In *Windows on the Japanese Past: Studies in Archaeology and Prehistory,* edited by Richard Pearson, pp. 11–56. Center for Japanese Studies, University of Michigan, Ann Arbor.

Tsukada, M., Shinya Sugita, and Yorko Tsukada
   1986   Oldest Primitive Agriculture and Vegetational Environments in Japan. *Nature* 322:632–634.

Ueyama, Shumpei
   1969   *Shōyōjurin Bunka: Nihon Bunka no Shinso (The Broad-Leaf Evergreen Zone Culture: Deep Structure of the Japanese Culture).* Chuokoronsha, Tōkyō.

Watanabe, Hitoshi
   1986   Community Habitation and Food Gathering in Prehistoric Japan: an Ethnographic Interpretation of the Archaeological Evidence. In *Windows on the Japanese Past: Studies in Archaeology and Prehistory,* edited by Richard Pearson, pp. 229–254. Center for Japanese Studies, University of Michigan, Ann Arbor.

Watson, Patty Jo
   1976   In Pursuit of Prehistoric Subsistence. *Midcontinental Journal of Archaeology* 1:77–100.
   1989   Early Plant Cultivation in the Eastern Woodlands of Eastern North America. In *Foraging and Farming,* edited by D.R. Harris and G.C. Hillman, pp. 556–571. Unwin Hyman, London.

Yasuda, Y.
   1978   *Prehistoric Environment in Japan: Palynological Approach.* Institute of Geography, Faculty of Science, Tōhoku University, Sendai.

Yoshizaki, M.
   1984   *Kokogaku ni okeru Ezo to Ezo-chi 1: Kodai Ezo-chi no Bunka* (The Ezo and Ezo Region from an Archaeological Perspective 1: the Ancient Ezo Culture). *Sozo no Sekai* 49:80–97.

Zvelebil, Marek
   1986   Mesolithic Societies and the Transition to Farming: Problems of Time, Scale and Organisation. In *Hunters in Transition,* edited by Marek Zvelebil, pp. 167–188. Cambridge University Press, London.

10

Non-Agricultural Staples and Agricultural Supplements: Early Formative Subsistence in the Soconusco Region, Mexico

Michael Blake and Brian S. Chisholm
University of British Columbia

John E. Clark
Brigham Young University

Karen Mudar
University of Michigan

The more we learn about the transitions to agriculture, the more apparent it is that differences from region to region, and even within single culture areas, can no longer be glossed over and ignored. In developing models that try to explain the beginnings of agriculture, we often search first for similarities among all known cases, and then use the perceived similarities to construct universal models. It then becomes difficult to explain cases that do not fit the general pattern, and difficult to see just how they might be relevant to our understanding of the transition.

In a recent study, McCorriston and Hole (1991:47) examine the question of how plants were domesticated in the Levant and aptly state that :

> rather than promote a universal theory of the origins of agriculture on the assumption that the events worldwide share a universal explanation, we would prefer at this stage to see each region elucidated in its own terms, allowing us to consider all variables rather than only those held in common by all regions in which agriculture arose.

As Flannery (1986:5) has pointed out, this approach is useful because it achieves a high degree of "resolution," but at the cost of "perspective." A universal perspective can also be valuable, but only if based on sound case studies where we can have a high degree of confidence in the available data.

In the study reported here we attempt to answer the broad question: How pervasive was the transition to agriculture in Formative Mesoamerica? More particularly, we wish to know when and why did this transition take place along the Pacific coast of Chiapas, Mexico? And finally, what might this case tell us about some of the more general processes in the transition to agriculture? We examine these questions in light of two contrasting theories for the origins of agriculture.

Two Perspectives on the Transition to Agriculture

In recent years two distinctly different sets of theories have been offered to explain the domestication of plants and the origins of agricultural production. One theory emphasizes the productive potential of domesticated plants, particularly with respect to "...crisis in production, population growth, and dispersal" (Rindos 1984:280). Similarly, Flannery (1986:16,27) has suggested that Archaic hunters and gatherers may have begun to manipulate the planting of species whose availability would otherwise have been seasonally or annually unpredictable. The eventual domestication of plants might be seen as the unexpected outcome of generations of such interventions, leading to the selection of characteristics that increased productivity and reliability. Most such plants would have been an essential part of the subsistence system, but some, gourds for example, may have been important as tools.

Recently, Hayden (1990; this volume) has proposed a contrasting model, suggesting that agriculture originated in regions of plenty and was linked with emerging sociopolitical complexity. He observes that societies with big men, chiefs, or accumulators, would have originated in relatively rich and productive environments. They would have encouraged the production of plant and animal resources that could be harnessed for the production of surpluses used in feasting and aggrandizement. Most of the first species domesticated, he argues, were not staples, at least in their initial stages of domestication. Rather, they were labor intensive crops that could have been used as delicacies in feasting and gift-giving.

It seems to us, however, that the interaction among humans and the plants and animals they exploited would have been so variable from region to region throughout the world that both of these sets of theories need to be examined and tested on a case by case basis (Harris 1977). As Gebauer and Price (this volume) point out, there are 20 or so explanations that have been used for the transition to agriculture and many of these might have taken place simultaneously in particular cases. The two broad perspectives considered here might be seen as opposite ends of a continuum which encompasses a range of different explanations for the transition. At the same time, one or the other might be rejected depending upon the available data for particular cases.

In the Natufian case in Southwest Asia, for example, increasing sedentism, the emergence of social inequality, and the particular suite of environmental factors outlined by McCorriston and Hole, may all have been responsible for the increased reliance on harvesting grain and its domestication. However, in other instances, such as the case of Archaic Oaxaca in the highlands of Mexico described by Flannery (1986), emerging sociopolitical inequality seems unlikely to have played a significant role. Furthermore, as Flannery (1973:273) discusses, both of these cases of the origins of seed-crop cultivation in relatively arid

regions might have been distinctly different from agricultural origins in the humid tropical lowland regions of the world. Much more detailed work needs to be done to document the transitions to agriculture in the humid tropics where seed-crops, root-crops, and tree-crops were integrated into complex systems of agricultural production, along with the continued harvesting of wild plants and animals (Harris 1977).

Agriculture and Competition: Origins vs. Spread

The origin of agriculture is a somewhat different question and may involve different processes from the spread of agriculture. Many of the dynamic relationships between cultivation and sociopolitical processes that Hayden (1990, this volume) discusses might more profitably be explained in cases of the spread or diffusion of agriculture, not of its origins per se. It is important to distinguish between these two processes because they affect our perception of the history of domestication vs. the history of use of a particular plant (Rindos 1984).

Plants may come under initial domestication in one region and then, later, these genetically altered plants may spread into neighboring regions for a number of reasons, some of which might be similar to those for initial domestication and some of which might be quite different. The new uses of cultivated plants as they radiate out from their initial homeland might range from primary subsistence staples in one subsistence system, to occasional supplements in another. In either case, the spread of domesticates outside of their naturally occurring habitats was probably facilitated by both because cultivation allowed somewhat greater predictability and control over the annual variations of food availability, and because in some cases cultivation allowed higher levels of production (Wills 1988, this volume). This latter factor would have been especially important in situations of emerging sociopolitical inequality. For example, individuals in emerging complex societies might readily adopt systems of agricultural production, particularly if it improved their chances of accumulating surpluses for use in feasting and status competitions as well as for provisioning larger households. However, this must be considered a secondary process, following one where long-term and almost imperceptible changes took place in plant productivity during the initial domestication process.

One might also expect there to have been rapid shifts in the rate of spread and adoption of certain domesticates as they crossed thresholds of productivity. Once genetic changes proceeded to the point where nutrient yields, per unit of labor or per unit of land cultivated, surpassed that which could have been obtained from harvesting naturally available foods, people took advantage of agricultural production, even if it meant restructuring social and political relationships.

Food production in zones of scarcity

In environmental zones of scarcity, or regions of large seasonal fluctuations in productivity, or in areas subject to periodic long-term drought, the adoption of agricultural staples (especially ones at or near the threshold of productivity) was likely to have been closely related to buffering natural shortfalls (see Close and Wendorf, this volume). This process is similar to the one leading to initial domestication in the first place, but simply extends it geographically. This pathway for the spread of agriculture would explain why it was adopted in regions far from its initial core area and in social and political contexts that, contra Hayden (1990), had egalitarian organizations and, in some cases, maintained egalitarian social and political structures well up to the historic period. New World examples include some Puebloan peoples the southwestern U.S. (Eggan 1950), Iroquoian-speaking peoples in the Northeastern U.S. and eastern Canada (Trigger 1990), and many of the Tupian tribes in the Amazon basin (Steward 1948). Although there are exceptions, e.g., the Tupian groups, most of these examples are characterized by seasonal extremes in temperatures or rainfall (Wills 1988, this volume; cf. Plog 1990).

Food production in zones of plenty

In contrast, agricultural production may have spread into zones of plenty, i.e., regions with high natural productivity and relatively little seasonal fluctuation in availability, for three different reasons. The first (Hayden 1990, this volume) would have been the incorporation of labor intensive agricultural production into existing subsistence systems geared to achieving surpluses, for social and political deployment by aggrandizers (Clark and Blake 1992). This process would only have been likely if the cultigens were rare delicacies given away in public displays. These foods might be restricted to, or at least more important for, higher status members of the community.

A second reason would have been the adoption of labor-efficient staples that had already passed the threshold of marginal productivity. Depending on the availability of productive land, crops such as these might be difficult for the elites to control, and they might not have been initially used by aggrandizers in political competition. We would expect such agricul-

tural crops to become rapidly incorporated into the diet of most people within a community.

The third reason for the adoption of agriculture in regions of natural abundance would be the use of cultigens by lower status individuals or even marginal communities within the society, who were disenfranchised with respect to natural resources. They may have been forced to invest in labor-intensive agricultural production to make up for reduced access to natural produce, especially during times of periodic shortfalls (see Hayden, this volume). This scenario presupposes an already emerging sociopolitical inequality with marginalized members of the community. In this case, we would expect that agricultural staples may not become generally used until late in a sequence and after the time that they are so productive that it makes little sense to rely entirely on wild food resources.

In effect, agricultural products might initially be seen as supplements, much as Flannery has described their initial role during the early stages of domestication. Non-agricultural resources would have been the dietary staples, even if agricultural products were known and technically available.

Early Formative Agriculture in Mesoamerica

Agricultural production has long been thought of as one of the cornerstones of early sedentary village life in Mesoamerica and one of the necessary conditions for the development of complex society. In fact, maize is usually seen as the primary staple crop in Mesoamerican prehistory from Formative times onwards. Its role is considered to be central to our understanding of Mesoamerican food production systems. MacNeish's work in the Tehuacán Valley has shown that the origins of maize and its integration into a system of agricultural production that included a variety of plants began as early as 7000 B.C. The earliest people to use and domesticate these plants were not sedentary, instead, they were nomadic foragers who incorporated these domesticates into a complex seasonal pattern of hunting and collecting (MacNeish 1967, 1972; Flannery 1968; Flannery 1986). However, in highland regions such as the Valley of Oaxaca and the Tehuacán Valley, by the time the first sedentary villages formed, hunting and collecting had to fit into an agricultural schedule that emphasized the planting and harvesting of maize, beans, and squash (Flannery 1973). Investigations in both areas indicate that early villagers probably did rely heavily on agricultural products, even though non-agricultural foods were still important (and continued to be so right up to the time of the Spanish Conquest).

A recent re-analysis by Farnsworth et al. (1985) of archaeological data from the Tehuacán Valley, including a stable carbon and nitrogen analysis of the human skeletal remains, suggests that a heavy dependence on grains, including maize, began as early as the Coxcatlán phase (ca. 5000–3000 B.C.). In Oaxaca, excavated macrobotanical remains show that domesticates, including maize, beans, squash, and avocados, were in use and consumed both before and after the appearance of the first sedentary villages (Flannery 1976, 1986). Kirkby's (1973) study of agricultural production suggests that the main staple, maize, was cultivated and relied upon from the Early Formative Tierras Largas phase (1400–1150 B.C.) onwards. She suggests, however, that maize did not reach a threshold of productivity, until about 1000 B.C. when larger varieties allowed greater yields per cultivated hectare of land. The assumption is that as maize cob size grew, and the plant became more productive, then early villagers came increasingly to rely on it as a subsistence staple. Both the Tehuacán and the Oaxaca data suggest that after agricultural products, particularly maize, became important in the subsistence system by the Late Archaic period, the trend towards increasing reliance on these plants continued through time.

The situation is not quite so clear for the lowland regions of Mesoamerica, partly because so few Archaic period sites have been excavated, and partly because subsistence remains have been recovered only sporadically from Formative period settlements. In the Gulf Coast lowlands, excavations at the Early Formative site of San Lorenzo Tenochtitlán, with sedentary occupations dating back to approximately 1500 B.C., recovered no direct evidence of agriculturally produced plant foods (Coe and Diehl 1980:137). At nearby La Venta and surrounding sites, recent excavations reported by Rust and Sharer (1988) have revealed macrobotanical remains of domesticated plants, including maize and beans. However, the state of preservation in this humid tropical zone is much poorer than in the arid highlands, and it is difficult to determine the ubiquity and importance of domesticates in the assemblage. There are ample faunal remains in the Gulf Coast lowlands to suggest that fishing and hunting were very important and that Early Formative peoples must have relied heavily on these wild resources, even if they did practice agriculture.

Along the Pacific coast of southeastern Mesoamerica, relatively few sites have yielded macrobotanical evidence of cultigens among their subsistence remains. Until several new archaeological projects began in the 1980s, Coe and Flannery (1967) were almost the only researchers to report domesticates at Early Formative sites along the Pacific Coast of either Chiapas or Guatemala. At the site of Salinas La Blanca they found mineralized casts of maize cobs, stalks, and leaves in Cuadros phase deposits, dating to approximately 1000–900 B.C. (Coe and Flannery 1967:71-73). However, for the five or six centuries of Early Formative occupation prior to the Cuadros phase, no cultigens had been reported. For example, excavations at La Victoria (Coe 1961), Altamira (Green and Lowe 1967), and Paso de la Amada (Ceja Tenorio 1985), all failed to recover any plant remains. Therefore, in terms of changes in the relative importance of agriculture from 1500 to 1000 B.C., archaeologists such as Michael Coe (1961:115), discussing the Ocós phase at the site of La Victoria on the Guatemalan coast, could only state the following: "The evidence of corn or any other kind of agriculture is purely inferential."

More recently, discoveries in the Mazatán region have recovered the charred remains of maize, beans, and other plants in deposits as early as the Locona phase (1350–1250 B.C.) (Clark et al. 1987; Blake et al. 1992a). These findings parallel those of Rust and Sharer (1988) in the La Venta region, but in neither the Gulf coast nor the Pacific coast do we have the same detailed paleobotanical information that is available from the arid highlands. We can now show that coastal folk, like their highland neighbors, knew of agriculture and even grew the standard Mesoamerican domesticates—but how important were these domesticates in the diet and how did the use of domesticates change through time? Several important questions remain: how did the transition to agriculture, especially maize cultivation, take place? How did maize, the staple of later periods, make its way into the diet of Early Formative people in the coastal regions of Mesoamerica? Were domesticates, such as maize, staples during this early period or were they merely supplements? Were they incorporated as part of expanding systems of sociopolitical inequality, or were they regular parts of the diet consumed by people regardless of emerging status differentiation? We will try to take some initial steps towards answering these questions by examining the subsistence data recovered from Early Formative sites near the modern town of Mazatán in the Soconusco region, located along the Pacific coast of Chiapas, Mexico.

Mazatán Region

Archaeological remains left behind by Early Formative villagers in the Mazatán region of Pacific coastal Mexico provide data relevant to answering some of the questions raised above. As we hope to demonstrate, these data also shed some light on the theoretical issues raised earlier, i.e., how and why did people living in zones of plenty make the transition to agriculture? Recent excavations and survey in the region have contributed a great deal of new information about some of Mesoamerica's earliest sedentary villagers (beginning about 1550 B.C.), and we can now begin to discuss their changing subsistence practices and sociopolitical organization.

Here we will briefly summarize: (1) The key aspects of the natural environment of the Mazatán region, especially as they relate to subsistence strategies, (2) changes in the sociopolitical character of Early Formative society in the region, beginning with the Barra phase at 1550 B.C., and (3) the evidence for Early Formative subsistence practices and relative importance of agricultural vs. wild produce.

Natural Environment of the Mazatán Region

The Mazatán region lies on the broad coastal plain of southeastern Mexico, hemmed in by the straight shoreline of the Pacific Ocean on one side and the Sierra Madre Mountains on the other (Figure 1). Once

one descends from the piedmont zone, there is very little topographic relief. The low coastal plain at elevations below 30 m (above sea level) extends inland between 15 and 20 km from the ocean.

Seasonal variation in temperature is minimal—the mean monthly temperature remains above 20°C all year—but there is a marked rainy season lasting from early May until late October. Precipitation increases towards the mountains, with the zone nearest the ocean receiving the least rainfall. Precipitation also varies considerably on a northwest-southeast axis along the coast, with the southern part of the coast receiving the most rainfall. This altitudinal and latitudinal variation in rainfall has a profound impact on the density and distribution of plant communities, and thereby animal communities, all along the coast of the Soconusco. Their distributions are highly heterogeneous and there is a great deal of variation in the suitability of sub-regions for hunting, collecting, fishing, and farming. The flora is dense; most of the piedmont zone was covered with a thick tropical

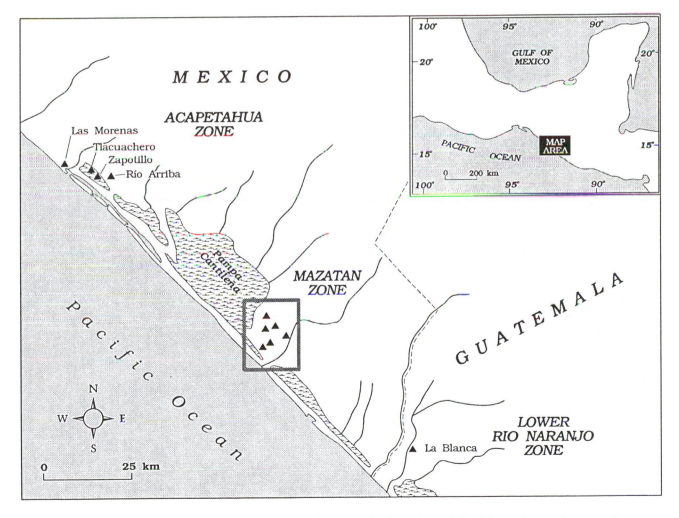

Figure 1. Map of the Soconusco region, showing the location of the Mazatán study zone (box outline—see Figure 2) and the archaeological sites mentioned in the text.

evergreen forest which is now only visible in small remnant parcels that have not been cut for modern agriculture. The coastal plain was originally covered with sub-tropical deciduous forest and savanna, but it too has been largely cleared. There are now also many open areas along the edges of the numerous rivers and swamps within the region which formerly had a dense, tropical, riparian forest.

The most variable habitats occur in the estuaries and swamps where there are both fresh and brackish water communities. Terrestrial and aquatic animal species are abundant and there are no marked seasonal variations in these, with the exception of some species of migratory waterfowl and a few species of anadromous fish.

In general, there is a good deal of diversity of both plant and animal life, especially in and near the rivers and estuaries, and there is little seasonal variation in the availability of most species. Compared with some of the more arid highland regions of Mesoamerica, there is also little year to year variation in the avail-

ability of plant and animal species that could have been exploited by prehistoric inhabitants of the region. At the risk of reducing the natural environment to an absurd level of simplicity, it is, comparatively speaking, a zone of plenty. Today the Soconusco region is considered to be one of the most agriculturally productive regions in Mexico (Helbig 1964), and various food crops can be grown year-round, even without irrigation. The Soconusco is also renowned for its hunting and fishing.

Early Formative Sociopolitical Organization

In several other papers we have outlined the evidence that Early Formative peoples in the Mazatán region had begun the process of the transition from egalitarian to rank society, beginning during the Barra phase (1550–1350 B.C.) and accelerating during the succeeding Locona phase (1350–1250 B.C.) (Clark and Blake 1992; Blake and Clark 1991). Settlement patterns and associated obsidian data show a two-tiered settlement hierarchy by the Locona phase, and the region

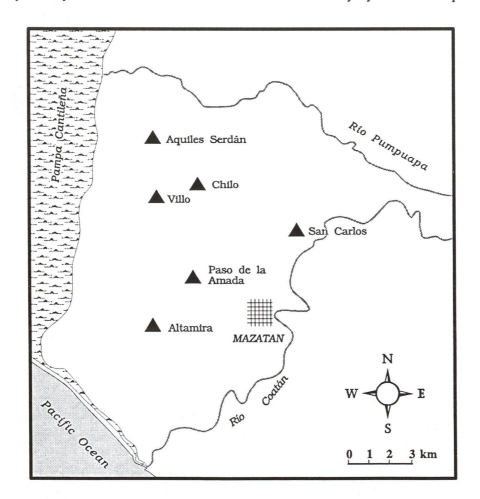

Figure 2. Principal Early Formative period settlements in the Mazatán zone. Aquiles Serdán, Chilo, Paso de la Amada, and San Carlos may all have been the centers of small emergent chiefdoms by the Locona phase (1350–1250 B.C.) and each was surrounded by numerous smaller hamlets (not shown).

may have been divided into at least four separate political units (Figure 2) (Clark and Salcedo 1989). In the largest villages within each polity, there were elevated mounds that preserve evidence of substantial architecture. Buildings such as Structure 4 at the site of Paso de la Amada, show a considerable labor investment and long-term continuity in their location and building style (Blake 1991; Blake et al. 1992b). Structures such as these may have been chiefs' residences, but there remains the possibility that they served public functions as well (Figure 3).

Artifact assemblages associated with Barra and Locona phase villages, particularly the ceramics, suggest emerging sociopolitical inequality. All of the

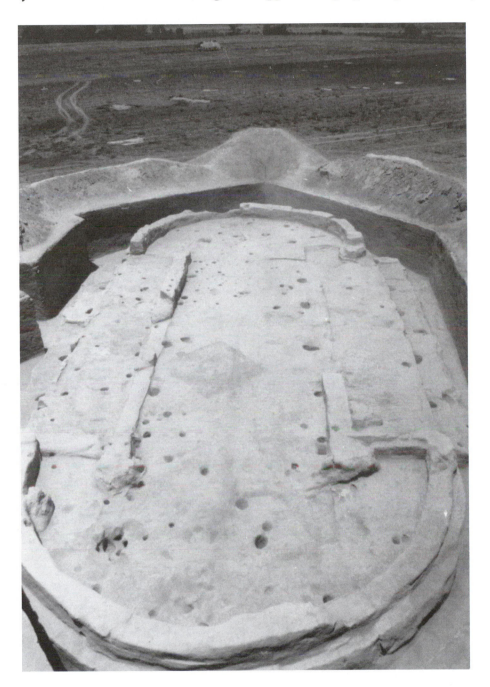

Figure 3. Structure 4, Mound 6, Paso de la Amada. This structure was built and occupied during the Locona phase and may have been a chief's residence. Visible on the structure's platform are standing semi-circular and straight clay "walls" or benches. The long axis of the structure measures 22 m. Small fish bones and other refuse were found scattered about the floor (Blake et al. 1991b).

Barra and Locona phase ceramic vessels consist of slipped and decorated containers that probably had food-serving functions (Clark and Blake 1990). None of the Barra phase ceramics and few of the Locona phase vessels appear to have been used for cooking or storage (Figure 4). Some flat-bottomed dishes were also made of stone, representing a huge investment of skilled labor.

Although we have little in the way of mortuary data with which to examine the question of status differentiation, the combination of settlement patterns, distribution of trade goods such as obsidian, architectural information, and ceramics and lapidary, all tend to indicate the same thing. Beginning during the Barra phase, and continuing through the Locona, Ocós (1250–1150 B.C.), and Cherla (1150–1000 B.C.) phases, permanent and possibly hereditary sociopolitical inequities emerged. It is in this context that we will now turn to a brief examination of the subsistence remains and an assessment of the importance of agriculture for these early villagers.

Subsistence Data

Faunal Remains. Animal bones constitute one of our most abundant sources of subsistence information.

Analysis of the faunal remains has been begun by Kent Flannery and Karen Mudar, using comparative collections at the University of Michigan. Although analysis is still underway, a complete analysis has been carried out for a Cherla phase refuse pit at Aquiles Serdán (Flannery and Mudar 1991). The following discussion summarizes their preliminary report which will eventually be included in the final site report.

The refuse pit consists of a shallow depression, 3 m in diameter and 60 cm deep, dug into an abandoned Ocós phase house floor (Blake et al. 1992b: Figure 14). It contained a "fine fraction" of smaller faunal remains, primarily fish bones, that likely accumulated from sweeping the patio of a nearby Cherla phase house. The "coarse fraction" was largely absent, and must have been discarded elsewhere at the site. Other excavation proveniences had greater proportions of larger bones, indicating that the pit does not necessarily represent the site-wide proportions of one species to another. However, it probably does give a good picture of the relative proportions of the species of fish and other small animals consumed.

Flannery and Mudar found that the faunal assemblage suggested a heavy use of freshwater estuary fish. The most common identifiable species were

Figure 4. A reconstruction drawing of Locona phase ceramic vessels from several sites in the Mazatán region. Dark stippling represents polished red slip, light-on-dark stippling represents irridescent pink on polished red slip, and white represents unslipped. The large dish in the foreground (lower left) is estimated to have a rim diameter of approximately 25 cm (other vessels are drawn to approximately the same scale).

represented by more than 30,000 bone elements and included *mojarra negra (Cichlasoma trimaculatum)*, gar (*Lepisosteus tropicus*), and catfish (possibly *Arius* sp.). Table 1 shows the identified fish species present along with the estimated minimum number of individuals (MNI). Their relative proportions, by MNI count, are shown in Figure 5. The individuals represented in the excavated deposits occurred in a wide range of sizes. Mojarra grow up to 30 cm while the gar and catfish grow up to 1 m. The excavated assemblage included large quantities of small individuals probably taken by net, as well as many large individuals that might have been taken by hook. In confirmation of this, we have recovered Locona and Ocós phase ceramic net weights, as well as Cherla phase bone fish hooks (Figure 6).

Reptiles were also well represented in the Aquiles Serdán refuse pit and those identified by Flannery and Mudar include turtles, crocodiles and/or caymans, iguanas, and snakes (Table 1). The two most common species of turtles were the small *casquito* or mud turtle (*Kinosternon* cf. *cruentatum*) and the jicotea or pond slider (*Pseudemys* cf. *grayi*). They also identified three possible fragments of *Dermatemys*, a large river turtle that is not native to the region and may have been traded in as a turtle shell drum.

Since the land mammal and bird remains found in the Aquiles Serdán refuse pit were not as well represented as elsewhere at the site or at other sites in the region, only brief mention will be made of them (Table 2). They include a few individuals each of white-tailed deer (*Odocoileus virginianus*), domestic dog (*Canis familiaris*), pocket gophers (*Orthogeomys grandis*), collared peccary (*Dicotyles tajacu*), raccoon (*Procyon lotor*), gray squirrel (*Sciurus aureogaster*), armadillo (*Dasypus novemcinctus*), cottontail rabbit (*Sylvilagus floridanus*), and opossum (*Didelphis virginianus*). Bird species represented included: grebe (*Podiceps nigricollis*), darter (*Anhinga anhinga*), green heron (*Butorides virascens*), tiger-heron (*Tigrisoma mexicana*), goose (*Anser* cf. *albifrons*), coot (*Fulica americana*), and duck (*Anas* spp. and *Cairina moschata*). There were also several smaller species of birds, perhaps taken for their feathers: short-tailed hawk (*Buteo* cf. *brachyurus*), a smaller hawk (*Accipiter* sp.), and an unidentified parrot, a jay, and another songbird.

Deposits from sites of all other Early Formative phases have also yielded large quantities of faunal remains but the analysis of most proveniences is still underway. Fish remains are the most common by frequency at all the other sites spanning the Barra to Cherla phases. For example, the floor of Locona phase Structure 4 had dozens of small fish bones and gar scales embedded in it and few larger bones from other animals. Like the Cherla phase refuse pit at Aquiles Serdán, this suggests regular floor sweeping

with only the smallest debris, including bones, left behind.

At some sites located nearer the ocean, such as the Ocós phase component at Sandoval, the most common faunal remains consisted of shells of brackish water mollusks (Clark et al. 1987). This, plus the limited range of ceramic vessels when compared with those found at other inland Ocós phase sites, suggests that the site's residents specialized in harvesting shellfish, and may only have resided there seasonally. Further inter-site comparisons await the complete faunal analysis from all of the excavated sites.

Plant Remains. Macrobotanical remains were also relatively common in the well preserved deposits, but so far, not enough of them have been identified to give us a clear picture of the range of plant species exploited. As with the faunal analysis, the plant identifications are currently underway and definitive statements about the taxa present and their relative ubiquity must await the completion of this study. However, early on in our sorting of the flotation samples, we recovered charred maize cupules, cob fragments, and charred beans. The earliest of these

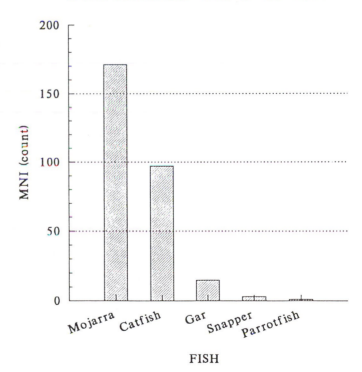

Figure 5. Bar graph showing the estimated frequencies of fish types (MNI) identified in the Cherla phase (1150–1000 B.C.) refuse pit at Aquiles Serdán (see text for scientific names of genera and species). Based on data from Flannery and Mudar (1991).

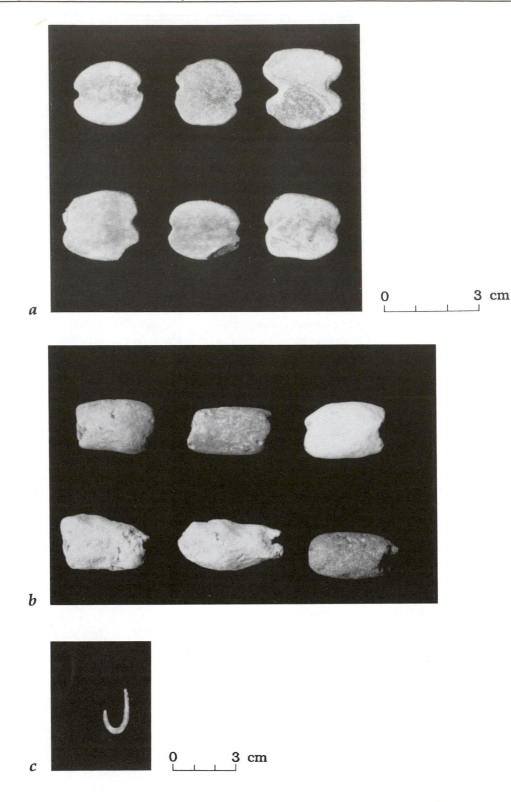

Figure 6. Ceramic net weights and a bone fish hook from Early Formative sites in the Mazatán region. (a) notched sherd weights from Locona phase deposits; (b) crude "bead"-shaped ceramic weights from Ocós phase deposits; (c) bone fish hook from Cherla phase deposits at Aquiles Serdán.

Table 1. Fish and reptile taxa by genus, species or family and estimated minimum number of individuals (MNI) in the Cherla phase (1150 -1000 B.C.) refuse pit at Aquiles Serdán (after Flannery and Mudar 1991).

Taxa	Common Name	Estimate MNI	Habitat
FISH			
Cichlasoma trimaculatum	mojarra negra	171	freshwater and brackish
Arius sp.	catfish	97	marine and tidewater rivers
Lepisosteus tropicus	gar	15	freshwater
Lutjanidae	snappers	3	marine
Scaridae	parrotfish	1	marine
REPTILES			
Kinosternon cf. *cruentatum*	mud turtle	several	freshwater
Pseudemys cf. *grayi*	pond slider turtle	several	freshwater
Dermatemys sp. (poss.)	river turtle	rare	freshwater
Crocodylus sp.	crocodile	1	freshwater
Crocodilian (unidentified)	crocodile and/or cayman	several	freshwater
Ctenosaura sp.	black iguana	several	terrestrial
Iguana iguana	green iguana	several	arboreal
Boa (unidentified)	Boa snake	several	terrestrial and aquatic
Snakes (unidentified)	various	several	unknown

Table 2. Mammal and bird taxa by genus and species and estimated minimum number of individuals (MNI) in the Cherla phase refuse pit at Aquiles Serdán (after Flannery and Mudar 1991).

Taxa	Common Name	Estimated MNI
MAMMALS		
Odocoileus virginianus	white-tailed deer	5
Canis familiaris	domestic dog	4
Orthogeomys grandis	pocket gopher	9
Dicotyles tajacu	collared peccary	1
Procyon lotor	raccoon	1
Sciurus aureogaster	gray squirrel	1
Dasypus novemcinctus	armadillo	1
Sylvilagus floridanus	cottontail rabbit	2
Didelphis virginianus	opossum	2
BIRDS		
Podiceps nigricollis	grebe	1
Anhinga anhinga	darter	1
Butorides virascens	green heron	1
Tigrisoma mexicana	tiger-heron	1
Anser cf. *albifrons*	goose	1
Fulica americana	coot	2
Anas spp.	ducks	several
Cairina moschata	muscovy duck	1
Buteo cf. *brachyurus*	short-tailed hawk	1
Accipiter sp.	small hawk	1
unidentified	parrot	1
unidentified	jay	1
Meleagris gallopavo	wild turkey	1

were found in Locona phase refuse and ash deposits at the site of Chilo, so we know that by at least 1350 B.C. two important Mesoamerican cultigens were being grown and consumed in the Mazatán region. At present, no cultigens have been identified in Barra phase deposits, likely due to the small sample of plant remains. Cultigens, such as maize and beans, do seem to be more common in later deposits, particularly the Cherla and Cuadros components, than in the earlier Locona and Ocós phase deposits.

Other plant remains are also present in the excavated Early Formative deposits. We recovered several examples of large charred seeds that resemble avocado seeds, but these have not yet been identified with reference to comparative collections. Several types of smaller seeds, all as yet unidentified, were also recovered from the flotation and heavy fraction sample. In spite of the fact that plant remains, including cultigens, have been recovered from the excavations, there are few methods available for determining their relative importance in the subsistence system and in the diet of the prehistoric population as a whole. With respect to cultigens, the question remains: how important were these plants in the diet?

The previous work by Coe and Flannery (1967) at Salinas la Blanca, 30 km to the southeast along the coast, would suggest that by the Cuadros phase, cultigens such as maize were becoming important subsistence crops. The cultivation of maize was clearly not started on the coast by Cuadros phase peoples, and Coe and Flannery (1967:104) suggest that it may have been underway "...as far back as the Ocós phase..." Our findings in the Mazatán region confirm that both Ocós (1250–1150 B.C.) and Locona (1350–1250 B.C.) phase peoples grew maize and beans. Perhaps we will find that Barra phase people did as well. If maize was a significant part of the diet of Locona and Ocós phase villagers in the Mazatán region, then there should be other evidence of its importance. One potential line of evidence comes from stable carbon isotope analysis, a technique that has proven useful for assessing changes in the importance of maize in the diet of prehistoric peoples in the New World.

Stable Carbon Isotope Analysis of Human Bone. Stable carbon isotope analysis has recently been used with a great deal of success in determining the changing importance of maize in the diet of prehistoric peoples in the New World (Spielmann et al. 1990; Matson and Chisholm 1991; Farnsworth et al. 1985; Ericson et al. 1989; Bender et al. 1981; van der Merwe and Vogel 1978). The basic premise that allows an assessment of changing maize consumption by measuring the ratio of $^{13}C/^{12}C$ (expressed as $\delta^{13}C$) in human bone collagen is that maize is one of a few New World crops, known as C4 plants, that produce relatively high $\delta^{13}C$ values.

The C4 plants, use a pathway (Hatch-Slack) for photosynthesis that incorporates CO_2 carbon into a four carbon molecule. These include mainly xeric grass species, such as "... maize, some millets, some sorghums, cane sugar, some amaranths, and some chenopods" (Chisholm 1989:12). The more common C3 cycle plants (Calvin-Benson) include most of the flowering plants, trees, shrubs, and most of the temperate zone grasses (Chisholm 1989:12) and produce relatively low $\delta^{13}C$ values. Another more rare group of plants, known as CAM (Crassulacean Acid Metabo-lism), are made up of tropical succulents, such as pineapple and various cacti (Chisholm 1989:12), and also yield high $\delta^{13}C$ values. Like C4 and CAM plants, marine organisms also have carbon pathways that lead to higher $\delta^{13}C$ values and their consumption can affect the $\delta^{13}C$ measurements in humans who rely on marine foods.

By measuring the $\delta^{13}C$ values of prehistoric human bone, we can estimate the relative importance of C4 and/or CAM plants in the diet compared with C3 plants (either direct or indirect consumption). Our expectations are as follows: low $\delta^{13}C$ values indicate a diet high in C3 plants and/or animals that consume C3 plants, and high $\delta^{13}C$ values indicate a diet high in C4 or CAM plants and/or animals that consume C4 or CAM plants. High $\delta^{13}C$ values can also be produced by consumption of marine foods. While the results of such an analysis taken by themselves are not conclusive, when combined with other lines of evidence, they can lead to reliable interpretations. For example, low $\delta^{13}C$ values would rule out the possibility of there having been significant amounts of C4 or CAM plants, either directly or indirectly, in the diet. Therefore, with lower $\delta^{13}C$ values, maize and any other C4 or CAM plants could not have been consumed regularly in large quantities. Conversely, with higher $\delta^{13}C$ values, C4 and/or CAM plants or marine foods must have been consumed in larger quantities and on a more regular basis.[1]

We have recently concluded a preliminary isotopic analysis of 29 samples of human bone collagen from several sites in the Soconusco region, spanning the Late Archaic to the Late Postclassic period (Blake et al. 1992a). Figure 7 shows the distribution of average $\delta^{13}C$ values by phase or period and by sub-region. The sub-regions are Acapetahua, in the Chantuto estuary system 50 km to the northwest of Mazatán, and Río Naranjo, also in an estuary setting 30 km to the southeast of Mazatán (Figure 1). We will first discuss the implications of the 15 Early Formative period samples from sites in the Mazatán sub-region. Then we will compare these results with those from the adjacent sub-regions in the Soconusco and $\delta^{13}C$ values from other parts of the New World.

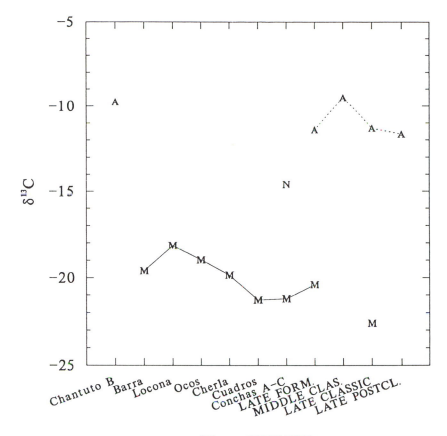

Figure 7. Average $\delta^{13}C$ values for each phase and PERIOD, by geographic sub-region. (A) Acapetahua; (M) Mazatán; (N) Lower Río Naranjo. Based on data presented in Blake et al. (1991a).

The $\delta^{13}C$ values for the Mazatán Early Formative samples range from -22.6 to -16.9 and have an average of -19.4 (Table 3). Figure 8 presents box plots of the Mazatán Early Formative $\delta^{13}C$ values by phase. The box plots present the median and range of $\delta^{13}C$ values by phase and show that, during the course of the period, there was no trend toward increasing reliance on C4 or CAM plants. As pointed out elsewhere (Blake et al. 1992a), these plants, including maize, could not have made a significant contribution to the diet of any of these individuals. The average is as low as that measured for hunting and gathering groups in Northeastern North America and Northern Europe, and similar to values recorded for Europeans subsisting on C3 plants and terrestrial animals (Chisholm 1989; Vogel and van der Merwe 1977).

The trend during the Early Formative period in the Mazatán zone is one of continued non-reliance on maize, a pattern that even extends into later periods for those few individuals so far recovered (Figure 7). In contrast, samples recovered by Barbara Voorhies in the neighboring Acapetahua sub-region and by Michael Love in the Lower Río Naranjo sub-region, indicate

that by the Middle Formative period there was a rapid shift to increased reliance on C4 or CAM plants (Blake et al. 1992a). Figure 7 shows where the crucial gaps in our sequence exist and why the diachronic and regional trends must still be considered tentative. In order to refine our understanding of both regional and chronological variation in the $\delta^{13}C$ values, we need more samples from the Early and Middle Formative periods in the Acapetahua sub-region, from all but the Conchas A-C phases in the Lower Río Naranjo sub-region, and from the Late Archaic and the Classic and Postclassic periods in the Mazatán sub-region.

One period in particular that we would like to see better represented is the Late Archaic period. Figure 7 shows that in the Acapetahua zone the average $\delta^{13}C$ values (from two samples from the site of Tlacuachero) for the Chantuto B phase (2700–1800 B.C.) is -9.6. This average is as high as those for the later, presumably maize-using periods in the Acapetahua sub-region. If these two Chantuto B phase individuals had such high $\delta^{13}C$ values because they were C4 or CAM plant consumers, then the possibility exists that there was a substantial shift in diet between the Late Archaic and

Table 3. $\delta^{13}C$ values for Early Formative period human bone samples from the Mazatán sub-region (after Blake et al. 1991).

Site	Lab Id. Number	$\delta^{13}C$ Value
Cuadros Phase		
Aquiles Serdán	1675	-21.2
Aquiles Serdán	1686	-22.4
Villo	1681	-20.2
Cherla Phase		
Paso de la Amada	1685	-21.5
Aquiles Serdán	1643	-18.2
Ocós Phase		
Paso de la Amada	1676	-18.7
Paso de la Amada	1678	-19.2
Paso de la Amada	1680	-18.4
Aquiles Serdan	1679	-19.6
Locona Phase		
Chilo	1644	-19.6
Chilo	1645	-16.9
Chilo	1669	-18.7
Chilo	1672	-17.4
Barra Phase		
San Carlos	1683	-20.5
Paso de la Amada	1687	-18.7
Average	n=15	-19.4

the Early Formative. The shift represents a significant decline in C4 or CAM plant consumption. If maize consumption was a major cause of this high average $\delta^{13}C$ value for the Chantuto B phase individuals, then it must have sharply declined by the beginning of the Early Formative period, at least in the Mazatán sub-region, where C4 and CAM plants were not a significant part of the diet.

Marine food consumption was not the likely cause of higher $\delta^{13}C$ values in the two Archaic individuals from Tlacuachero. An analysis of the stable nitrogen isotopes ($\delta^{15}N$) showed that the two Tlacuachero samples were not enriched compared with several individuals from later sites in both the Acapetahua and Lower Río Naranjo zones (Blake et al. 1992a).[2] Furthermore, there was little evidence of marine food consumption in the faunal remains from Tlacuachero (Voorhies 1976), although Voorhies, Michaels, and Riser (1991) hypothesize that shrimp were harvested

and processed at the site. We do not know how shrimp consumption would affect the stable carbon and nitrogen values.

The possibility remains, therefore, that some Late Archaic inhabitants of the Soconusco consumed (directly or indirectly) C4 and/or CAM plants at about the same level as did inhabitants of the region after about 800 B.C., until the time of the Spanish Conquest. However, during the millennium between the end of the Late Archaic period and the beginning of the Middle Formative period, C4 and/or CAM plants, while present, could not have been a significant part of the diet. The isotope data suggest that they declined in significance just as people began to settle the region in permanent villages by about 1500–1600 B.C. We will now examine the implications of this pattern for our understanding of the transitions to agriculture and the evolution of settled village life in this part of Mesoamerica.

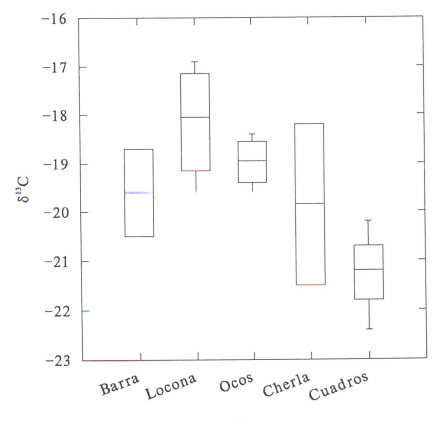

Figure 8. Box plot of median $\delta^{13}C$ values and ranges, by phase, for the Mazatán region (Table 3). The plots show that $\delta^{13}C$ values do not increase through time.

Discussion and Conclusion

Previous models for the evolution of settled village life and the role of agriculture in that transition led us to expect that we would find abundant archaeological evidence that Early Formative villagers in the Soconusco were agriculturalists. In particular we expected to find that they cultivated maize, beans, and other domesticated plants that were commonly used in Mesoamerica. In our excavations of well preserved Ocós and Cherla phase deposits at the site of Aquiles Serdán, we found such dense faunal material, primarily fish bone, that we initially thought we might be dealing with fishers and hunter-gatherers. However, in our initial flotation samples from the very same deposits, we began to recover small numbers of maize cupules, other seeds, and beans, and eventually carbonized cob fragments. This confirmed our initial expectations that these early villagers were agriculturalists and we had found the first concrete evidence of pre-Cuadros phase cultigens in the Soconusco.

It became clear that these Early Formative villagers combined fishing and hunting with at least some reliance on agricultural production and the consump-tion of those products. However, the question remained: what was the relative importance of these subsistence regimes and how did they change through time? Further, were any observable changes in these correlated with changing sociopolitical complexity?

The preliminary results of the stable carbon isotope analysis indicate that the *presence* of a plant in a pale-obotanical assemblage does not necessarily mean that it was an important part of the diet. We find macrob-otanical evidence of maize, for example, at some of the Early Formative sites we have excavated, but the stable carbon isotope analysis shows that all 15 Early Formative individuals from those sites consumed much less maize than Middle Formative and later people in neighboring regions. Although they certainly did eat maize, along with beans and other cultivated and wild plants, the faunal and isotopic evidence suggests that estuarine resources were prob-ably more important. If the isotope analysis is correct, then C4 plant use during the Early Formative period was so low that the stable carbon isotope values in bone collagen are indistinguishable from those recorded for hunting and gathering groups in other parts of the New World and Europe.

Our data allow us to suggest the following patterns: (1) Early Formative villagers in the Mazatán region relied heavily on fishing and hunting, and probably the gathering of wild plants. (2) They also grew (or perhaps imported) cultigens including maize and beans. However, maize and other C4 plants were not dietarily significant. (3) Through the Early Formative there is no evidence that maize consumption increased. Even well beyond the Early Formative, some individuals in the Mazatán sub-region continued to rely on many foods except maize. However, our initial findings suggest that in the regions adjacent to Mazatán, by the Middle Formative period, people began to rely more heavily on maize as a food resource. This suggests regional differences in production and consumption strategies by the Middle Formative which continued at least until the Late Classic period: some people in the rich estuary zone in the Mazatán region continued to subsist primarily on fishing and hunting, while their neighbors began to cultivate and consume increasing amounts of maize.

The Mazatán Early Formative case suggests that there may be distinctly different trajectories for both the origins and spread of agriculture in different parts of the world and within ecologically diverse culture areas such as Mesoamerica. In zones of relative scarcity, or unpredictability, agriculture may develop as both Flannery (1986) and McCorriston and Hole (1991) point out, as a means of buffering against seasonal shortfalls and even responding to prolonged environmental deterioration.

The extension of agriculture into zones of relative scarcity may proceed as a way of moving predictable resources to environments where they do not naturally occur as a way of continuing the buffering process. One need not posit competitive feasting as a driving mechanism for such a change.

Conversely, in zones where natural abundance is great to begin with and where sedentary complex societies could have arisen without much reliance on agricultural production, we must look for a different explanation for the adoption of agriculture. In Hayden's (1990 and this volume) scenario, the adoption of agriculture is regarded as a supplement to naturally occurring resources; agriculture was really just an extension of the manipulation of food stuffs in order to increase productivity. In such situations one might expect agriculture to become quickly incorporated into societies where there is evidence for social complexity and where such mechanisms as competitive feasting and gift giving require increased production.

However, as we have seen in the Mazatán case, this is not necessarily so, at least in the sense of adopting agriculture for the production of staple resources. Our data show that complex societies might occasionally adopt agricultural products as supplements while maintaining a reliance on naturally occurring staples, such as fish, game, and wild plants for a long period.

The transitions to agriculture, both world-wide, and within Mesoamerica, must have been as varied as the environments and peoples who occupied them. Our general understanding of the evolutionary processes that led to the transition to agriculture and, conversely, of the role that agriculture has played in the processes of cultural evolution must rest on many well-documented prehistoric cases from around the world. We do not yet consider the Soconusco case to merit the designation "well-documented." However, enough is now known about the region to allow us to identify the strengths as well as the weaknesses in our coverage of the archaeological record. In order to more clearly sort out geographic from diachronic variation, we need comparative samples from some of the hitherto unrepresented phases and periods within each of the sub-regions of the Soconusco. To compare the Mazatán sequence with the sequences in neighboring regions, we need to know more about the Early and Middle Formative periods in the Acapetahua region, more about the Archaic period in the Mazatán region, and more about the Archaic and the Late Formative periods in the Río Naranjo region. Until more subsistence and isotope data are available for these periods in the Soconusco, the question of the timing and causes of the transition to agriculture will remain vague.

Acknowledgements

Our archaeological fieldwork and laboratory analysis was funded by the New World Archaeological Foundation, Brigham Young University, and the Social Sciences and Humanities Research Council of Canada. Kent Flannery and Karen Mudar analyzed the faunal remains from Aquiles Serdán. Brian Chisholm oversaw the isotopic analysis and was assisted by Michael Hanslip who helped prepare the human bone samples. Bente Nielsen, Department of Oceanography, University of British Columbia, carried out the mass spectrometry analysis. We thank Barbara Voorhies and Michael Love for making available bone specimens for isotopic analysis and for discussing the results of their own work at great length. The Consejo de Arqueología of the Instituto Nacional de Antropología e Historia in Mexico granted us permission to carry out our archaeological fieldwork in Chiapas and permitted the analysis of selected archaeological bone and plant samples in the U.S. and Canada. All archaeological materials from our excavations in Chiapas remain the property of I.N.A.H. and we thank them for permission to reproduce our photos of the excavations and artifacts. Comments on and discussions about this paper have contributed greatly to our understanding of the issues involved, and for that we thank: Barbara Voorhies, Jim Brown, Gary Feinman, Brian Hayden, Denise Hodges, Joyce Marcus, and Gil Stein. In particular, though, we would like to thank Kent Flannery who, in commenting on an earlier version, helped us to find a "Sunsweet Prune" pit or two.

Notes

[1] In the present case, both CAM plants (which are rare in the humid tropics [Chisholm 1989]) and marine foods (such as marine mammals, fish, and shellfish which are rarely found in the Early Formative archaeological deposits in the Mazatán region) would probably not have contributed significantly to high $\delta^{13}C$ values.

[2] Schoeninger and DeNiro (1984) and Schoeninger (1989) provide clear explanation of the process of stable nitrogen isotope fractionation in marine organisms and their consumers.

References Cited

Bender, M. M., D. A. Baerreis, and R. L. Steventon
  1981 Further Light on Carbon Isotopes and Hopewell Agriculture. *American Antiquity* 46:346-353.
Blake, M.
  1991 An Emerging Early Formative Chiefdom at Paso de la Amada, Chiapas, Mexico. In *The Formation of Complex Society in Southeastern Mesoamerica*, edited by W. L. Fowler, Jr., pp. 27-46. CRC Press, Boca Raton.
Blake, M., and J. E. Clark
  1992 The Emergence of Hereditary Inequality; the Case of Pacific Coastal Chiapas, Mexico. In *The Evolution of Archaic and Formative Society Along the Pacific Coast of Latin America*, edited by M. Blake. Washington State University Press, Pullman, in press.
Blake, M., B.S. Chisholm, J. E. Clark, B. Voorhies and M. W. Love
  1992a Early Farming, Fishing, and Hunting along the Pacific Coast of Mexico and Guatemala. *Current Anthropology*, in press.
Blake, M., J. E. Clark, V. Feddema, M. Ryan and R. Lesure
  1992b Early Formative Architecture at Paso de la Amada, Chiapas, Mexico. *Latin American Antiquity*. in press.
Ceja Tenorio, J. F.
  1985 Paso de la Amada: An Early Preclassic Site in the Soconusco, Chiapas. *Papers of the New World Archaeological Foundation*, 49. Brigham Young University, Provo.
Chisholm, B. S.
  1989 Variation in Diet Reconstructions Based on Stable Carbon Isotopic Evidence. In *The Chemistry of Prehistoric Human Bone*, edited by T. D. Price, pp. 10-37. Cambridge University Press, Cambridge.
Chisholm, B. S., D. E. Nelson, and H. P. Schwarcz
  1982 Stable-carbon Isotope Ratios as a Measure of Marine Versus Terrestrial Protein in Ancient Diets. *Science* 216:1131-1132.
Clark, J. E., and M. Blake
  1990 The Development of Early Formative Ceramics in the Soconusco, Chiapas, Mexico. Paper presented at the 55th annual meetings of the Society for American Archaeology, Las Vegas, Nevada, April 18-22.
  1991 The Power of Prestige: Competitive Generosity and the Emergence of Rank Societies in Lowland Mesoamerica. In *Factional Competition and Political Development in the New World*, edited by E. M. Brumfiel and J. W. Fox. Cambridge University Press, Cambridge, in press.
Clark, J. E., and T. Salcedo
  1989 Ocós Obsidian Distribution in Chiapas, Mexico. In *New Frontiers in the Archaeology of the Pacific Coast of Southern Mesoamerica*, edited by F. Bove and L. Heller, pp. 15-24. Arizona State University Press, Tempe.

Clark, J. E., M. Blake, P. Guzzy, M. Cuevas and T. Salcedo
  1987 El Preclásico Temprano en la Costa del Pacifico: Informe Final. Ms. on file, Instituto Nacional de Antropología e Historia, Mexico, D.F.
Coe, M. D.
  1961 La Victoria, an Early Site on the Pacific Coast of Guatemala. *Papers of the Peabody Museum of Archaeology and Ethnology*, No. 53. Harvard University, Cambridge.
Coe, M. D., and R. A. Diehl
  1980 *The People of the River. In the Land of the Olmec*, vol. 2. University of Texas Press, Austin.
Coe, M. D., and K. V. Flannery
  1967 Early Cultures and Human Ecology in South Coastal Guatemala. *Smithsonian Contributions to Anthropology*, vol. 3. Smithsonian Institution, Washington, D.C.
Eggan, F.
  1950 *Social Organization of the Western Pueblos*. University of Chicago Press, Chicago.
Farnsworth, P., J. E. Brady, M. J. DeNiro, and R. S. MacNeish
  1985 A Re-evaluation of the Isotopic and Archaeological Reconstructions of Diet in the Tehuacán Valley. *American Antiquity* 50:102-116.
Flannery, K. V.
  1968 Archaeological Systems Theory and Early Mesoamerica. In *Anthropological Archeology in the Americas*, edited by B. J. Meggers, pp. 67-87. Anthropological Society of Washington, Washington, D.C.
  1973 The Origins of Agriculture. *Annual Review of Anthropology* 2:271-310.
  1976 Empirical Determination of Site Catchments in Oaxaca and Tehuacán. In *The Early Mesoamerican Village*, edited by K. V. Flannery, pp. 103-117. Academic Press, New York.
  1986 *Guilá Naquitz*. Academic Press, Orlando.
Flannery, K. V., and K. Mudar
  1991 Feature 6 of Aquiles Serdán. Ms. on file, Department of Anthropology and Sociology, University of British Columbia, Vancouver.
Green, D. F., and G. W. Lowe (editors)
  1967 Altamira and Padre Piedra, Early Preclassic Sites in Chiapas, Mexico. *Papers of the New World Archaeological Foundation*, 20. Brigham Young University, Provo.
Hayden, B.
  1990 Nimrods, Piscators, Pluckers, and Planters: The Emergence of Food Production. *Journal of Anthropological Archaeology* 9:31-69.
Helbig, K. M.
  1964 La Cuenca Superior del Río Grijalva, Un Estudio Regional de Chiapas, Sureste de México. Translated by Felix Heyne. Instituto de Ciencias y Artes de Chiapas, Tuxtla Gutiérrez, Chiapas.
Kirkby, A. V. T.
  1973 The Use of Land and Water Resources in the Past and Present Valley of Oaxaca. In *Prehistory and Human Ecology of the Valley of Oaxaca*, Vol. 1, edited by K. V. Flannery. Memoirs of the University of Michigan, Museum of Anthropology 5, Ann Arbor.
MacNeish, R. S.
  1967 A Summary of the Subsistence. In *The Prehistory of the Tehuacán Valley, Vol. 1: Environment and Subsistence*, edited by D. S. Byers, pp. 290-309. University of Texas Press, Austin.
  1972 The Evolution of Community Patterns in the Tehuacán Valley of Mexico and Speculations About the Cultural Processes. In *Man, Settlement, and Urbanism*, edited by P. J. Ucko, R. Tringham and G. W. Dimbleby, pp. 67-93. Gerald Duckworth and Co., London.
Matson, R. G., and B. S. Chisholm
  1991 Basketmaker II Subsistence: Carbon Isotopes and Other Dietary Indicators from Cedar Mesa, Utah. *American Antiquity* 56(3): 444-459.
McCorriston, J., and F. Hole
  1991 The Ecology of Seasonal Stress and the Origins of Agriculture in the Near East. *American Anthropologist* 93(1):46-69.

Plog, S.

1990 Agriculture, Sedentism, and Environment in the Evolution of Political Systems. In *The Evolution of Political Systems*, edited by S. Upham, pp. 177-199. School of American Research Book, Cambridge University Press, Cambridge.

Rindos, D.

1984 *The Origins of Agriculture: An Evolutionary Perspective*. Academic Press, New York.

Rust, W. F., and R. J. Sharer

1988 Olmec Settlement Data from La Venta, Tabasco, Mexico. *Science* 242:102-104.

Schoeninger, M. J.

1989 Reconstructing Prehistoric Human Diet. In *The Chemistry of Prehistoric Human Bone*, edited by T. D. Price, pp. 38-67. Cambridge University Press, Cambridge.

Schoeninger, M. J., and M. J. DeNiro

1984 Nitrogen and Carbon Isotopic Composition of Bone Collagen in Terrestrial and Marine Vertebrates. *Geochimica et Cosmochimica Acta* 48:625-639.

Spielmann, K. A., M. J. Schoeninger, and K. Moore

1990 Plains-Pueblo Interdependence and Human Diet at Pecos Pueblo, New Mexico. *American Antiquity* 55:745-765.

Steward, J. H.

1948 Culture areas of the Tropical Forests. In *The Tropical Forest Tribes*, edited by J. H. Steward, pp. 883-899. Handbook of South American Indians, vol. 3. Smithsonian Institution Bureau of American Ethnology, Bulletin 143, Washington, D.C.

Trigger, B.G.

1990 Maintaining Economic Equality in Opposition to Complexity: an Iroquoian Case Study. In *The Evolution of Political Systems*, edited by S. Upham, pp.119-145. School of American Research Book, Cambridge University Press, Cambridge.

van der Merwe, N. J., and J. C. Vogel

1978 δ^{13}C Content of Human Collagen as a Measure of Prehistoric Diet in Woodland North America. *Nature* 276:815-816.

Vogel, J. C., and N. J. van der Merwe

1977 Isotopic Evidence for Early Maize Cultivation in New York State. *American Antiquity* 42:238-242.

Voorhies, B.

1976 The Chantuto People: An Archaic Period Society of the Chiapas Littoral, Mexico. *Papers of the New World Archaeological Foundation*, 41. Brigham Young University, Provo.

Voorhies, B., G. H. Michaels, and G. M. Riser

1991 Ancient Shrimp Fishery. *National Geographic Research & Exploration* 7(1):20-35.

Wills, W. H.

1988 *Early Prehistoric Agriculture in the American Southwest*. School of American Research Press, Santa Fe.

11

Plant Cultivation and the Evolution of Risk-Prone Economies in the Prehistoric American Southwest

W.H. Wills
University of New Mexico

The primary focus of this chapter is the role of plant cultivation in economic systems of the prehistoric American Southwest. Most studies of early agriculture in the Southwest, especially regarding the adoption of domesticated plants from Mesoamerica, have been mainly concerned with the accurate dating of initial cultigens and the patterning of early agricultural sites relative to particular environmental contexts. Concern here is with the affect of food production on indigenous economic strategies. I will argue that the first use of cultigens was an important tactic for enhancing the effectiveness of economic systems which were based fundamentally on hunting and gathering. The adoption of cultigens was a significant development, but it did not promote immediate changes in economic organization.

Although plant cultivation is a consistent part of the archaeological record through time following the first acquisition of domesticates, there are few data that directly indicate any effort to intensify food production for over a thousand years. Moreover, I will suggest that even after the establishment of settlements with substantial architectural structures—thought to reflect increased sedentism and agricultural reliance—the evidence for *change* in the basic organization of economic production is often ambiguous. In my view, the first qualitative changes in early southwestern mixed economies are organizational features associated with risk-taking and increasing marginal returns.

Economic Change in the Late Archaic

Food production in the American Southwest began in the preceramic period with the introduction from Mesoamerica of a suite of three domesticated plants—maize, squash and beans—poorly adapted to the southwestern environment (Ford 1981; Galinat 1985; Benz 1986). Chronometric data indicate that domesticated plants were introduced to the Southwest region by 1200 B.C. and possibly a few hundred years earlier (Table 1). The oldest direct radiocarbon dates on cultigens come from archaeological sites in the lower Rio Grande Valley and the montane regions of west-central New Mexico and the Colorado Plateaus, suggesting that the Rio Grande Valley may have been the corridor for transmission of these new food resources. However, there is considerable overlap in the standard deviations associated with radiocarbon assays on domesticates from different regions. A conservative assessment of directional trends points to widespread use of cultigens by approximately 1000 B.C. In culture historical terms, incipient food production took place during the Late Archaic (ca. 1500 B.C. to A.D. 200), the preceramic period immediately preceding the adoption of pottery and an investment in substantial architectural structures (see Cordell 1984; Cordell and Gumerman 1989). Some researchers believe that immigrants may have brought cultigens into the region (e.g., Huckell 1990), but most archaeologists see an immediate commitment to new resources by indigenous foragers.

Table 1. Selected Radiocarbon Assays on Cultigens from Southwestern Archaeological Sites. [1]

Site	Date (BP)	Lab No.	Cultigen	Calibrated Date Range [2]
Tornillo	3175±240	GX-12720	*Zea mays*	1733 - 1112 BC
Bat Cave	3120±70	A-4188	*Zea mays*	1491 - 1320 BC
Bat Cave	3010±150	A-4167	*Zea mays*	1440 - 1015 BC
Bat Cave	2980±120	A-4186	*Curcubita pepo*	1377 - 1052 BC
Sheep Camp	2900±230	A-3388	*Curcubita pepo*	1430 - 830 BC
Three Fir	2880±140	Beta26271	*Zea mays*	1314 - 845 BC
Fairbank	2815±240	AA-4457	*Zea mays*	1373 - 790 BC
Milagro	2780±90	AA-1074	*Zea mays*	1074 - 830 BC

[1] For more inclusive lists of reported radiocarbon assays on prehistoric cultigens consult: Long et al. 1986; Simmons 1986; Wills 1988a, 1988b; Smiley and Parry 1989, 1990; Huckell 1990. Two outlier dates on *Zea mays* not included above are 3740±70 BP (A-4187) from Bat Cave and 3610±170 BP (?) from Three Fir Shelter; see the text for details on these two early assays.

[2] Calibrated calender year range (one sigma) using program CALIB. See Stuiver, M., and P.J. Reimer (1986), A Computer Program for Radiocarbon Calibration. *Radiocarbon* 28:1022-1030.

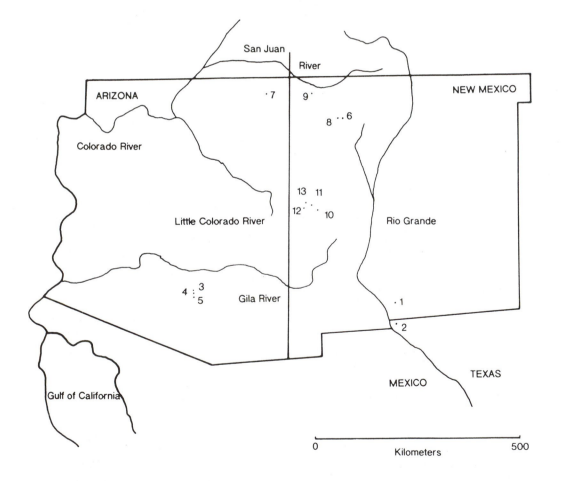

Figure 1. Locations of Archaeological Sites and Geographical Areas Mentioned in the Text. Lower Rio Grande Valley: (1) Tornillo Rockshelter, (2) Keystone Dam. Tucson Basin: (3) Fairbank, (4) Milagro, (5) AZ EE:2:30. Colorado Plateaus: (6) Sheep Camp Shelter, (7) Three Fir Shelter, (8) LA 18091, (9) Shabik'eshee Village. Mogollon Highlands: (10) Bat Cave, (11) Tularosa Cave, (12) O Block Cave, (13) SU site.

Early cultigens are found at sites in a wide range of topographic and environmental settings (Figure 1), but the availability of water was a common requirement for site location. Incipient cultivation sites found in the arid zones of southern Arizona are closely linked to piedmont riverine settings characterized by grasslands, oak woodlands, and high water tables (Huckell 1990). Archaeological sites in other parts of the Southwest with early cultigens occur at higher elevations (1500 to 2000 m), where precipitation was somewhat greater.

Hunter-gatherer populations in the Southwest prior to the adoption of cultigens had annual and seasonal mobility systems structured in large part by the spatial and temporal patterns of resource availability in seasonal arid climates. Most interpretations of pre-agriculturual settlement data infer seasonal movements of forager bands between winter/spring camps at lower elevations near streams and rivers and summer/fall camps at higher elevations in woodlands and forests (see Toll and Culley 1984; Huckell 1984, 1990; Shackley 1985; Wills 1988a; MacNeish 1989).

The Lower Rio Grande Valley

The oldest dated specimens of maize in the Southwest have been recovered from Tornillo Rockshelter, located in the Organ Mountains near Las Cruces, New Mexico. Upham et al. (1987) reported a date of 3175 ± 240 B.P. (GX-12720) on a combined sample of eight maize cob fragments. The presence of early maize in lower Rio Grande Valley is not surprising as this area should have been a primary corridor between Mesoamerica and the Southwest. Despite this early maize date, evidence for agricultural production in this area is spotty—at best—until sometime after about A.D. 900, about 2000 years later.

The Lower Rio Grande Valley courses through the Chihuahuan Desert, an extremely arid region without marked patterns of seasonality. North-south trending mountain ranges provide some zonation in habitats, and consequently some variation in wild resources, but the majority of economic food sources are desert species that are spatially heterogeneous but generally available throughout the year. Archaeological data from Archaic sites in this area indicate major procurement foci on succulents such as cacti and agave, as well as mesquite beans and grass seeds (Fields and Girard 1983). The introduction of maize around 1100 B.C. does not seem to have necessitated any alteration of basic foraging patterns, which apparently involved small and highly mobile bands (Carmichael 1986).

There is good evidence from the Keystone Dam locality, just four kilometers from the Rio Grande near El Paso, that Archaic groups in the Lower Rio Grande Valley established seasonal residential camps by 2000 B.C., where they built small houses, conducted intensive on-site plant processing activities, and may have dug storage pits (O'Laughlin 1980). Interpreted as a seasonal base-camp, the Keystone Dam locality indicates repetitive and extended site occupation well before the arrival of domesticated plants in the region. Moreover, the basic settlement pattern documented by 2000 B.C. for the region—consisting of riverine base-camps and short-term foraging sites away from drainages—apparently continued without fundamental change until A.D. 1000 (Fields and Girard 1983:39; Carmichael 1986:218; Seaman et al. 1988:139).

Almost all the direct evidence for agriculture in the Lower Rio Grande in the Late Archaic comes from rockshelters located in upland, montane areas. Recent excavations by MacNeish (1989) have recovered maize cobs from horizons assigned to the Late Archaic, but only the direct date on cobs from Tornillo Shelter has been reported. Plant cultivation throughout the Archaic in the Lower Rio Grande Valley was apparently confined to relatively wetter locations in the uplands occupied during spring and fall months (e.g., Carmichael 1986:11).

Evidence for a continuous presence of cultigens through the 2000 year period following introduction until the emergence of small villages is limited. However, assuming that domesticates were maintained within local economies throughout this time span, it is clear that they did not stimulate the emergence of agricultural economies. All known archaeological sites from this long interval consist of short-term open-air campsites or rockshelters. If plant husbandry provided a consistent dietary component during this period, then its economic role appears to have been the same for nearly two millenia.

Southeastern Arizona

Some of the best data on early agriculture in the Southwest have come from excavations of Late Archaic sites in the Sonoran desert grasslands of southeastern Arizona. Direct AMS dates on maize cupules from the Fairbank and Milagro sites in the Tucson area indicate cultivation by 800 B.C. (Huckell and Huckell 1988, 1990). Numerous direct dates on maize from other sites in the region provide substantive support for continuous involvement with cultigens throughout the Late Archaic after 800 B.C. (Long et al. 1986; Huckell 1990).

Early agricultural sites occur extensively in the broad river valleys draining montane regions in southeastern Arizona, especially near Tucson. Investigations at sites along the San Pedro River, the Tanque Verde and Santa Cruz rivers, and Cienega Creek have revealed a range of Late Archaic sites associated with maize macrofossils, often represented by middens exposed in deeply dissected alluvium (Sayles 1983; Eddy and Cooley 1983; Huckell 1984, 1990; Waters 1986; Roth 1989; Mabry 1990). Some of these sites have extremely high densities of artifacts, especially fire-cracked rock, and are associated with small structures, burials, and deep, bell-shaped pits. Maize macrofossils are nearly ubiquitous in flotation samples and counts frequently exceed those in flotation samples from ceramic period settlements in the same area (Fish et al. 1990). Midden thickness (sometimes exceeding 50 cm) and superimposed features indicate regular re-use of some site locations. Most sites occur along major stream channels where watertable farming would probably have been most effective (Huckell 1988, 1990).

The grasslands and oak woodlands that character-ized the intermontane valleys where large Late Archaic sites have been discovered offer a wide range of natural resources. Important economic plants such as mesquite, cactus, grasses, and agave occur in the valley bottoms and piedmont slopes, with acorns and pinyon found on mountain flanks. Mountain sheep, mule deer, and antelope were probably among the largest and most significant game animals, together with rabbits.

Huckell (1990:44) notes that vegetative diversity is high in this area because of the zonation associated with mountain ranges, and that distances between major habitats are not great, so that people could quickly gain access to a wide range of resources. He interprets riverine sites such as Milagro as seasonal basecamps associated with maize cultivation from which forays were mounted to acquire resources from nearby piedmont and mountain zones.

Huckell (1990) believes that large riverine Late Archaic sites indicate a major shift toward increased sedentism from the pre-agricultural period in this area. He states that Middle Archaic sites in the same area do not approach some of the Late Archaic agricultural sites in size, artifact density, burials, or presence of structures. Consequently Late Archaic populations represented "a substantially different socioeconomic system from the preceding Middle Archaic period" (Huckell 1988:74). It is possible these differences may be explained by the appearance of farming groups in the area, rather than the acquisition of domesticates by indigenous foragers (Huckell 1990:374).

Recent data from the Tucson area, however, may complicate the inference of dramatic social changes associated with the adoption of maize. For example, large pre-agricultural sites with high densities of arti-facts and features have been reported (Sayles 1983; Waters 1986; Huckell 1988:75). Mabry (1990) described a cultural horizon exposed in alluvium along the Santa Cruz River. The level was 80 m in length and lay strati-graphically beneath a Late Archaic pit house with a radiocarbon date of 3040±110 B.P. (Beta-39577). Flotation samples from the pithouse fill produced charred maize kernels and cob fragments and thus the radiocarbon date is the earliest yet associated with an agricultural site in the region. For comparison, the pre-pithouse use of the same location resulted in a cultural horizon as extensive as some reported with later agri-cultural sites (Mabry 1990:5). While data from the Middle Archaic are still ambiguous, Fish et al. (1990:79) and James (1990:27) argue that pre-agricul-tural adaptations in the Tucson area could have been relatively sedentary with groups occupying limited ranges, permitted by the high resource productivity of the region. Burials predating 4000 B.P. have been reli-ably reported (Waters 1986:52,64) and may occur earlier (see Sayles 1983:124 and Haury 1950), attesting to some degree of residential stability in the Middle Archaic. Given these data and Huckell's (1988:75) observation that many Middle Archaic sites in alluvial areas were probably eroded away during the Middle Holocene, it seems somewhat uncertain whether the intensive use of campsites found with early cultigens was qualitatively different from prior settlement patterns of pre-agriculturalists.

The rapid pace of Archaic period research in the Tucson area will probably resolve this issue in the near future. For the time being, the lack of formal variation among Late Archaic agricultural sites is striking. According to Huckell (pers. comm.) for example, the main change in settlement pattern over 800 years in the Cienega Creek Valley was shifting site locations attributed to changing levels in the local water table. Huckell (1990:365-370) also feels that Late Archaic sites throughout southeastern Arizona represent a consis-tent economic regime wherein basecamps associated with maize cultivation were seasonally abandoned during spring months to allow foraging for foods such as cactus and mesquite. Some of those sites have date ranges exceeding 200 years (Roth 1989; Huckell 1990), suggesting long-term, repeated use. Regardless of how different this strategy may have been from pre-agricul-tural periods, particularly in the intensity of site use, it seems apparent that once established, the settlement type associated with cultivation persisted for a mille-nium relatively unchanged.

Colorado Plateaus

Two sites in the basin and range country of the Colorado Plateaus have direct dates on cultigens of approximately 900–800 B.C. A squash seed dated at 2900±230 B.P. (A-3388) was recovered from Sheep Camp Shelter in Chaco Canyon, New Mexico, while a cob fragment from Three Fir Shelter on Black Mesa, northeastern Arizona, was assayed at 2880±140 B.P. (Beta-26271). Numerous AMS dates on maize and squash occur between 800 B.C. and A.D. 200 in northern Arizona and New Mexico, as well as southern Utah (Simmons 1986; O'Leary and Biella 1987; Smiley and Parry 1990). An AMS date of 3610±170 B.P. on maize from Three Fir Shelter has also been reported, although the principal investigators urge that this extreme outlier be considered cautiously in archaeological reconstructions. Site context for each of these earliest currently dated cultigens in the Colorado Plateaus are rockshelters that were probably used for convenient natural storage.

Open-air Late Archaic sites containing directly dated maize are seldom older than about 200 B.C. An early date for maize was obtained from LA 18091 in the San Juan Basin (Simmons 1986), although it has a large standard deviation that reduces its accuracy (2720 ± 265 B.P.; UGa4179). Rockshelters were utilized throughout the Colorado Plateaus during the Archaic, but during the early agricultural period there seems to have been greater investment in storage facilities and increasingly intensive utilization of the shelters. Structures interpreted as shallow pithouses do occur in the northern Rio Grande Valley as early as 5000 B.P. (Lang 1980:54). Some of these structures were found with large artifact inventories, including numerous grinding tools, as at the Moquino site, dating to 4600 B.P. (Beckett 1973).

Large buried alluvial sites with radiocarbon dates between 3300 and 2180 B.P. occur near Albuquerque and have characteristics similar to those found in dissected arroyos of southern Arizona (Agogino and Hibben 1957). Clearly, some evidence for relatively intensive site use in riverine settings is present before maize cultivation begins in this area. Nevertheless, small pit structures like those described by Huckell (e.g., 1990) and Mabry (1990) in the Sonoran Desert are not evident in the Late Archaic of the Colorado Plateaus and adjacent northern Rio Grande Valley until about 200 to 100 B.C. (Smiley 1985). As in the southern portions of the Southwest, the initiation of plant cultivation, while associated with greater use of natural storage locations, seems to have had little overall effect on settlement patterns for hundreds of years or more.

Mogollon Highlands

In the Mogollon Highlands of west-central New Mexico, radiocarbon dates from Bat Cave indicate cultivation of squash and maize as early as 1100 B.C. Three maize fragments have produced dates ranging from 3120±70 B.P. (A-4188) to 3010±150 B.P. (A-4167) and a squash seed dated to 2980±120 B.P. (A-4186) (see

Long et al. 1986; Wills 1988a). As at Three Fir Shelter on Black Mesa, Bat Cave produced a single anomalous early maize date (3740±170 B.P.; A-4187) that the principal investigators view with extreme caution (Wills 1988a:109). The archaeological data from Bat Cave and other nearby Late Archaic sites such as Tularosa Cave and O Block Cave suggest a settlement history similar to the Colorado Plateaus, with rockshelters offering natural storage and processing locations for early farmers. Open-air sites in the Mogollon Highlands with evidence for agriculture are not evident until A.D. 200–400 and may be associated with the first appearance of ceramics in the region (LeBlanc 1982; Wills 1988a, 1988b). As in the northern Southwest, rockshelters in the Mogollon Highlands were utilized by hunter-gatherer groups throughout the Archaic period, but the tempo and intensity of use increased after the adoption of cultigens.

Bat Cave provides the best current data for subsistence strategies associated with early maize cultivation in the highlands. Investigations in 1948-50 (Dick 1965) and 1981-83 (Wills 1988) identified complex stratigraphic relationships among artifacts and material that seem to indicate dramatic changes in site use with the appearance of cultigens in the shelter deposits. Human utilization of this high-elevation site (2021 m) began at least 10,000 years ago and the earliest dated cultigen is found at about 3100 B.P. During this long interval Bat Cave was used infrequently by small groups that left only small hearths and broken stone tools. The appearance of maize and squash at about 1100 B.C. corresponds to the first evidence at the site for storage pits and occupation floors formed by the compaction of ash and charcoal from numerous small firepits. Artifact density and diversity increases greatly during the pre-agricultural period. Hundreds of perishable items such as fragments of rabbit fur robes and yucca cordage are present. Radiocarbon dates indicate occupation throughout the Late Archaic, but superimposed occupation surfaces typically are separated by non-cultural deposits, pointing to episodic rather than continuous use. It is impossible to determine the duration of site use from the relative imprecision of the radiocarbon assays, but it appears that some of the non-occupation intervals may have been extensive, perhaps decades or longer.

The data from Bat Cave do not suggest a major reorganization of pre-agricultural systems in response to the acquisition of cultigens. Areas of the site that were used by people are actually fairly small and available evidence suggests that the role of the shelter during the early agricultural period was primarily for storage and processing. There are no indications of intensive occupation that might be the result of year-round residence.

The simple presence of cultigens at this high elevation site suggests a change in the seasonal movements characteristic of the pre-agricultural period. Cultivation of maize and squash requires planting in late spring or early summer. This is probably the least attractive season in upland areas of the Southwest; local resources are depleted or non-existent and many foods are available at lower elevations during this time (see Hevly 1983; Toll and Culley 1984). Maize cultivation at Bat Cave and other Late Archaic highland Mogollon sites clearly indicates a choice to be in resource poor locations, and to forego resources in other places.

I have argued that the relatively small rockshelter sites in the Mogollon highlands and the episodic occupation indicators in the Late Archaic do not reflect a major demographic shift to spring occupation or annual residency (Wills 1988a). Instead, I believe the evidence is consistent with small groups or bands moving into the upland zones somewhat earlier than in previous periods, and using stored maize as a temporary food source during these lean months. The objective of this earlier seasonal movement was probably not to cultivate maize, since its maturation in fall would occur at a time of tremendous abundance in wild foods. Late Archaic maize was not very productive (Galinat 1985) and should not have been competitive with foods such as pinyon, walnuts, sunflowers, and large game during autumn. Thus the cultivation of maize in highland locations seems tied to its storage potential and delayed use.

If that delayed use was not associated with year-round occupation, then the motivation for use in spring months seems linked to a desire or need to be in this area prior to the availability of wild foods. I have suggested that this is probably a monitoring tactic whereby groups normally moving seasonally from lower to higher elevation zones might send out small parties in advance of most of the population to assess the distribution and abundance of key resources that would not be available until the fall (Wills 1988a:147). Provided this interpretation is correct, it means that the use of domesticates in the highlands was geared toward increasing the efficacy of what was primarily a foraging economy.

Direct Evidence for Intensification of Food Production in the Late Archaic

Late Archaic archaeological features indicative of more intensive site use are consistent with mobility constraints imposed by plant cultivation. Domesticated plants are immobile resources requiring human attention; any use of domesticates should therefore be associated with prolonged use of particular localities.

But does that mean that agricultural dependence is the reason for those patterns? Obviously, use of the evidence for increasing occupational intensity as confirmation that sedentism occurs in order to accommodate increased agricultural production would be a circular argument. Instead, other independently derived data are needed. Such data can be found in three realms: (1) the technology associated with cultigen husbandry and processing, (2) bone chemistry indicators of dietary input, and (3) genetic patterns in plants resulting from cultivation.

One striking pattern is the absence of evidence for fields systems or water control in any region of the Southwest during the Late Archaic. Researchers have posited that food production during this period was confined to small gardens, possibly in localities with high water tables (Ford 1981,1984; Huckell 1990). The technology of water management does not appear archaeologically until about A.D. 700 in southern Arizona in the form of irrigation canals and is absent in other regions until perhaps the 11th century A.D. (Vivian 1970; Woosely 1980; Crown 1990). There may have been small features designed to enhance garden or field productivity during the Late Archaic, but there are no known formal field systems.

Several recent quantitative studies of grinding stone technology have examined tool morphology as a possible indicator of reliance on agriculture (Lancaster 1983; Adams 1990; Hard 1990; LeBlanc 1982; Martin and Rinaldo 1950). These studies begin with the assumption, supported by middle range research, that the surface area available for grinding increases with amount of time spent processing maize kernels. By inference, the size of hand stones (manos) is assumed to covary with dietary reliance on maize. However, evidence for agricultural dependence even in the early ceramic periods, 1500 years after the adoption of maize, is mixed. Hard (1990:143) found that manos from the Early Pithouse period (ca. A.D. 200–500) in the Mogollon area reflected patterns consistent with ethnographic groups having only minor reliance on agriculture. Morris (1990) found no conclusive evidence for temporal trends in Archaic grinding technology. In sum, grinding technology during the Archaic reflects generalized plant processing, not specialized function expected with heavy consumption of corn.

One of the more interesting negative pieces of evidence is the absence of ceramic vessels in the Late Archaic. Although the absence of pottery containers defines the Archaic period, there are fired-clay figurines from the Milagro site near Tucson (Huckell 1990:238), indicating that Archaic people were familiar with the basic principles of ceramic technology by 800 B.C. Moreover, ceramics were well-developed in

Mesoamerica by 1000 B.C., so it is curious that pottery did not move into the Southwest with agriculture. It is possible that mobility remained quite high in the Southwest, inhibiting the use of fragile containers, or that consumption of maize and beans was not significant enough to warrant the use of ceramic containers for boiling or steeping.

Bone chemistry studies have recently been cited as a direct method for assessing dietary intake from maize in the prehistoric Southwest (Katzenberg and Kelly 1988; MacNeish 1989). Matson and Chisholm (1986) presented evidence for $^{13}C/^{12}C$ ratios in human bone obtained from Late Archaic (ca. 200 B.C. to A.D. 100) sites in southeastern Utah. The values cited by Matson and Chisholm (1986) are among the highest reported for any archaeological sample anywhere in the world. They view these data as evidence for massive dietary contributions from maize. While this evidence indicates a high degree of dietary input from C4 plants and/or animals that consume large quantities of C4 plants, it is much less clear that this means maize. Indigenous C4 plants are among the most common economic plants recovered from southwestern archaeological sites. Wild C4 plants such as chenopodium are assumed to have constituted a major portion of pre-agricultural diets (Reinhard et al. 1985). CAM plants, which can exhibit C4 pathways, include extremely important economic species such as yucca, catcus, and agave. A conservative interpretation of the existing bone chemistry data is that during the Late Archaic there is evidence for intensive use of relatively low-return resources, such as grasses and desert succulents, as well as animals such as rabbits and small mammals that feed extensively on C4 plants. Maize was part of this suite of resources, but bone chemistry cannot yet demonstrate that maize was a major contributor (Schoeninger 1990).

The same conclusions can be drawn from existing data on phenotypic variation in maize during the Late Archaic. Maize possesses exceptional genetic plasticity. It is assumed that the extraordinary variety of maize in the New World reflects the relative ease with which humans can induce traits adaptive to varying edaphic, climatic, and topographic settings (Benz 1986: 22,381). For example, in the prehistoric Eastern United States, the introduction of maize around A.D. 200 was followed by rapid, culturally controlled changes in the plant (King 1987: 12). There are no indications of a similar pattern of rapid change in the maize first introduced to the prehistoric Southwest.

All early maize identified in the Southwest is either a primitive eight-row variety found throughout Latin America, or is 10 and 12 row Chapalote, the presumed derivative of the primitive eight-row maize (Ford 1981; Donaldson 1984; Galinat 1985, 1988; Adams 1990;

Huckell 1990). Unless more evolved maize varieties are recovered from Late Archaic sites, it would seem that for at least a thousand years (longer in some areas) the genetic potential of maize was not exploited by early food producers in the Southwest. Botanists have noted a similar lack of diversity in early southwestern cucurbits and beans (Cutler and Whitaker 1961:478; Kaplan 1981; Decker 1988). Consequently, the Late Archaic has so far produced little direct evidence for human manipulation of domesticates through phenotypic selection.

Apparently Late Archaic cultivators were not under pressure to increase the economic role of maize productivity through phenotypic change. In many ways this is surprising, since the primary economic advantage of maize cultivation is generally assumed to be its susceptibility to genetic control by humans for increased yields. One explanation may be that human population mobility was high enough so that there was little opportunity for effective isolation of maize populations or regular (i.e., annual) selection of seed crops. Whatever the answer, data now available do not seem to support the idea that maize was cultivated for increasing yields during the Late Archaic.

The lack of evidence for agricultural intensification—independent of increased occupation intensity at particular sites—contrasts with the widely held assumption that sedentism in the prehistoric Southwest came from dependence on agricultural surplus. Despite the presence of cultigens throughout the Late Archaic, direct information is lacking to indicate that food production led to fundamental changes in recipient economic systems.

Regional Summary

Current information from a rapidly growing number of investigated archaeological sites with early cultigens suggests that these plants were initially incorporated into forager economies without corresponding changes in the organization of subsistence strategies. It appears that the acquisition of cultigens was, in Redding's (1986) terms, a tactical shift in economic transformation. A lack of direct evidence for intensification of food production indicates that agricultural surplus was not the driving factor in the adoption process. Instead, indicators of overall economic intensification, including the bone chemistry evidence for low-return dietary items and more emphasis on storage, points toward a role for plant husbandry in facilitating increasingly localized economic pursuits.

Paleoclimatic data from the introduction period, between about 1500 and 1000 B.C., are not sufficient to establish any reliable causal links between environmental changes and subsistence strategies. Indicators such as pollen and patterns of alluvial deposition suggest that this period was probably wetter than at present and that there were no obvious episodes of dramatic climatic deterioration. The period of mid-Holocene elevated temperature and aridity known as the Altithermal (Hall 1985) may be most relevant to the adoption of domesticates in the southwest. This extensive climatic episode promoted widespread vegetation change that included the expansion of economically significant species such as piñyon and the development of grassland habitats that supported increased numbers of ungulates (see Van Devender et al. 1984; Wills 1988a). At a general level, these environmental developments likely stimulated seasonal patterns of mobility among hunter-gatherers that involved extended use of some localities during peak resource periods and may also have raised the regional carrying capacity for human populations. The adoption of domesticated plants during the Late Holocene is at least indirectly related to forager adaptations established in response to an environment increasingly advantageous to humans.

Economic Organization and Risk-Taking

Production Strategies

If plant cultivation during the Late Archaic contributed to economic intensification by making foraging tactics more effective, then economic organization during this period was not transformed by food production. Cultigens were not insignificant (Minnis 1985:310) but their importance was achieved through contribution to the maintenance of ongoing subsistence practices. This is not really surprising, as many researchers have posited that agriculture must be understood as one set of tactics systemically linked to a wide range of other economic and demographic factors (Flannery 1973, 1986; Green 1980; Hayden 1981, 1990; Redding 1986).

Eventually, however, there were changes in southwestern subsistence systems that led to large, sedentary communities. Few researchers doubt that agriculture, especially maize cultivation, played a significant and probably central role in the formation of large prehistoric villages and towns. By and large, most researchers also see the beginnings of a shift to agriculturally-dependent cultural systems taking place at the end of the Archaic period. In fact, the termination of the Archaic is identified by the appearance of pottery and settlements with multiple, substantial dwellings, which are presumed markers of increased agricultural dependence.

Many archaeologists seem to assume that the increased investment in site occupation (or "planning depth" in Binford's [1990] terms), indicates a different kind of economic system than was characteristic of preceramic periods. This is a useful question to pursue in assessing the role of food production: do ceramic period pithouse sites represent a fundamental change in economy from earlier time periods? I will distinguish here between two basic types of economic organization that generally characterize small-scale societies. One strategy may be described as "communal" and is commonly found among mobile hunter-gatherers. Communal systems are notable for strong emphases on generalized reciprocity or sharing among all members of a community, a limited degree of surplus production, and consequently a fairly poor capacity for economic intensification (Flannery 1972; Sahlins 1972; Lee 1990). In contrast, a "household" mode of production is characterized by restricted food sharing patterns (within individual households or between only a few households), some reliance on surplus production, and consequently a greater capacity for economic intensification (Byrd, this volume; Wilk and Netting 1984; Netting 1990).

The ability to intensify production is clearly advantageous to farmers who invest heavily in specific plots of land and consequently the differences between communal and household strategies are often discussed in terms of mobile foragers versus sedentary farmers (e.g., Winterhalder 1990). Logically, however, there is no necessary linkage between procurement strategies (e.g., hunting vs. plant husbandry) and the organization of production and consumption. A fairly extensive ethnographic literature reveals varying combinations of procurement activities and economic organizations (see Flannery 1972; Netting 1990). Archaeologists have an even larger number of prehistoric examples of complex social forms existing independently of agricultural production (e.g., Price and Brown 1985).

It may be more pertinent to think of communal and household organization in terms of productivity. The generalized reciprocity found in communal systems inhibits surplus production and consequently these strategies tend to be characterized by diminishing returns. A diminishing returns economy is one in which increased investment does not produce equal or greater returns. Social sanctions that promote sharing also inhibit individuals from producing more than they need to meet the demands placed upon them. Mobile hunter-gatherer societies typically experience diminishing returns as they move to a resource patch, utilize local resources until returns drop, and then move on to a new patch (Binford 1980; Bettinger 1980; Winterhalder 1981). Group mobility is probably an excellent indication of diminishing returns economies, but can also be found among pastoralists and in mixed economies as well as forager systems.

Economic systems among sedentary groups are likely to involve constant or increasing returns, since continued residence in a particular place requires equal or greater returns. Household organization allows individuals to keep surplus production and thus provides an incentive for intensification (Flannery 1972:42). Another way to distinquish these two kinds of organization in terms of productivity is by the kinds of feedback processes that support diminishing or increasing returns. Diminishing returns are the result of strong negative feedback influences; declining returns inhibit further investment in particular economic tactics. Increasing returns are typified by positive feedback since more investment results in better productivity (Arthur 1990).

Although these general distinctions are not equivalent, we may be able to learn more about the organization of production by considering contrasts between foraging and farming systems. For example, Winterhalder (1990) has argued that generalized versus restricted sharing patterns may be attributable to differences in the temporal intervals and spatial scales of production. In all small-scale societies it is probably correct to assume that food sharing is a risk management mechanism that reduces the variance in income experienced by individuals through averaging the incomes of all members of a groups (Hegmon 1989; Smith and Boyd 1990; Plog 1990). According to Winterhalder (1990:85), one problem with food sharing as a risk reduction tactic is individuals who "cheat" by not contributing as much as they receive. This problem is not difficult for hunter-gatherers to solve because their short production intervals last only hours or days and allow group members to accurately assess individual contributions. Farmers on the other hand have production intervals lasting months or years and cannot easily monitor individuals who shirk their obligations. For these reasons, Winterhalder (1990:86) maintains that farmers are better off restricting their sharing among closely-related individuals.

However, while restricted sharing helps solve the problem of cheating, it negatively affects the ability of individuals to reduce income variance by pooling the efforts and knowledge of a large number of people. Winterhalder (1990:84) suggests that farmers offset this loss by averaging the incomes of individual fields that may be widely dispersed over a local area. Field dispersal compensates for natural factors affecting productivity, such as spotty rainfall or predators. Dispersed fields may also be a means for allowing a small number of sedentary individuals to exploit a relatively large area, something that hunter-gatherers accomplish by mobility.

Of course, Winterhalder only suggests how variation in sharing can work for societies with different economies; he has not explained why one economic organization might be favored over another. Part of the explanation undoubtedly involves competition. As Netting (1990) points out, intensified land use seems to correlate with high population densities and resource scarcity. Under competitive conditions, the intensification potential that comes from restricted sharing and storage allows groups to control the use of particular locations and hence whatever resources occur in that place (Flannery 1972:47-49).

But competition alone cannot be a complete explanation for the adoption of household modes of production. Plog (1990) notes that agriculture has the potential to raise the productivity of an economy, but often does so at the cost of increasing variance around the mean output (see also Winterhalder 1990:72). Greater investment in cultivation means that catastrophic crop failure is more likely, due to simplified plant ecosystems (see Green 1980; Leonard 1989). In other words, surplus from food production can raise the mean income of an economic "individual" (say, for example, a family), but that increase may also be accompanied by greater variance.

Plog's argument is compatible with, and probably predictable from, general microeconomic theory. Sedentism requires an increasing returns economy and increasing returns economies are risk prone. Individuals in diminishing or constant returns systems avoid risk, that is they make decisions that will generate a small but certain return on a regular basis as opposed to less regular but larger returns. Diminishing returns economies typically have sigmoidal production curves (Figure 2A), while increasing returns are achieved by taking higher probabilities of loss in exchange for the occasional big payoff (see Baumol and Blinder 1986; Smith and Boyd 1990) and have logistic production curves (Figure 2B).

Therefore, household modes of production are fundamentally risk prone. The increased variance possible with the restriction of sharing to a small number of individuals or closely-related families may lead to higher mean levels of income but also leaves those individuals vulnerable to increased probability of loss. In order to understand why individuals would adopt a risk prone strategy, we need to examine the conditions in which such a strategy is beneficial.

Both theoretical and empirical data indicate that risk-taking is intimately linked to the distance between the mean income level and the minimum income level necessary for survival. Bartlett (1980:556–7) summarized a number of studies of risk-taking in agrarian

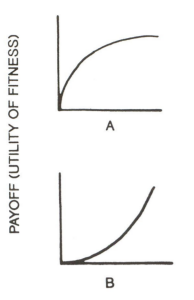

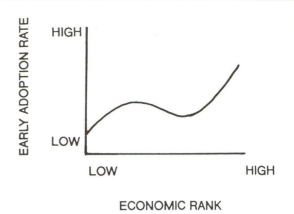

Figure 3. Empirical relationship between economic rank and the adoption of agricultural innovations proposed by Cancian (1979:3).

Figure 2. Payoff Functions and Risk Strategies (adapted from Smith and Boyd 1990: Figure 8.1). (A) Diminishing returns function associated with risk-averse strategies. (B) Increasing returns function associated with risk-seeking strategies.

societies (esp. Cancian 1979) and observed that the adoption of new resources, technologies, or production strategies tends to occur first and most rapidly among the wealthiest members of a community. Decisions to undertake innovations with a high degree of uncertainty (the outcome is unknown) or a high degree of risk (the outcome is associated with some probability of success) are initially made by individuals who can absorb losses if the innovation fails. Less wealthy members of a community wait to see if the innovation will work before adopting it, resulting in an adoption curve as seen in Figure 3.

Individuals whose income is well above minimum thresholds may adopt tactics that increase income variance because even if they experience losses, those declines may not be so large that income falls below the minimum threshold. Individuals whose mean income lies above the threshold but whose variance around the mean sometimes falls below the threshold will probably not adopt strategies that increase variance because it also increases the number of times they will experience shortfalls, unless the gain in the mean is sufficent that even higher variance still remains above the threshold. Hegmon (1989:93) follows Cancian (1979) in suggesting that individuals with mean incomes below the threshold will do anything to increase variance since that would take them above the threshold on at least some occasions, although this hypothetical situation seems unlikely in terms of agricultural production since individuals (or economies)

with less than minimum threshold incomes probably do not have the resources or residential stability required for extended production intervals.

The preceding observations about risk taking and income thresholds suggest that a necessary condition for restricted sharing and consequent enhancement of the potential for economic intensification of food production is a predictable and productive resource base. If increased variance in plant husbandry is the basis for achieving increasing returns, then the overall economy—i.e., the remaining foraging portion—must be adequate to maintain household organization in the event that crops fail. A decision to adopt household production strategies should be based on a perception that if cultivation fails, hunting and gathering tactics will provide adequate resources to sustain the system. In Cancian's (1979) terms, the initial adoption of new production strategies will be by individuals occupying or controlling a resource base that is "rich" relative to other individuals. Or, in Flannery's (1972:28) terms, "the origins of 'sedentary life' had more to do with the installation and maintenance of permanent facilities, and the establishment and maintenance of hereditary ownership of limited areas of high resource potential, than with agriculture per se."

Variation in the Economic Organization of Early Villages

Archaeological correlates of household modes of production can be drawn from cross-cultural studies by Flannery (1972), Binford (1990) and Netting (1990). First, restricted sharing should be evident in private food storage. Second, dwellings will be affiliated with household units, and often will be clustered rather than dispersed (see also Basehart 1973). And third, group composition will be fairly stable over time, often associated with land tenure, and may be evident in mortuary patterns such as cemeteries or burial in houses. The identification of these relationships archaeologically should be a good indication of household modes of production.

The earliest extensive evidence for plant cultivation occurs in the montane region of the central Southwest U.S., ca. 1200–1000 B.C., but there is no conclusive evidence for any major economic change until the beginning of the ceramic period, around A.D. 200-500, with the rather sudden appearance of substantial architecture at larger settlements. These settlements consist of semi-subterranean pithouses associated with numerous storage pits, burials, and high densities of artifacts. Sites with multiple dwellings are often called "villages" and this time period is generally thought to represent the first real commitment to settled communities in the Southwest, based largely on an increased dependence on agriculture (Lightfoot and Feinman 1982; Cordell 1984). Although Huckell (1990:362) argues that the Late Archaic pithouse sites near Tucson should also be considered villages, most researchers currently seem to feel that early ceramic settlements represent a qualitative increase in community size and permanence over the Late Archaic.

Temporal patterns of settlement use during the Late Archaic in the Mogollon Highlands can be described along an axis of occupational intensity defined as an ordinal scale (Figure 4). Numerical values were assigned to selected archaeological features that I believe reflect different degrees of settlement use. The presence of dwelling structures was assigned a value of 1, as I assume that structures indicate extended periods of site occupation. Storage pits were given a

value of 2 and indicate anticipated re-occupation of the locality. Burials received the highest value, 3, and are assumed to reflect groups attachment to the burial location, possibly as a symbol of ownership or land tenure. This valuation system is clearly simplistic and the literature contains numerous efforts to quantify more accurately the intensity of settlement use from archaeological data (e.g., Schiffer 1987; Nelson 1990; Powell 1990). Nonetheless, most of these studies have proven inconclusive (c.f., Schlanger 1990). The system proposed here has the advantage of allowing individual sites to be compared by single scores.

The resulting curve for the Mogollon Highlands (Figure 4) is logistic and reflects a rapid increase in occupational intensity after A.D. 200. To some extent, the intensity curve can be considered equivalent to the utility curves in Figure 2. The utility curves reflect the kinds of returns produced by an economy (or economic strategy) and we can assume, at least for heuristic purposes, that since occupational intensity is measured by features stemming from longer periods of site use, it is indirectly monitoring local economic productivity. If this equivalence is allowed, then the logistic curve for the Mogollon Highlands indicates a shift to risk prone production in the early ceramic period (compare with Figure 2).

The dramatic increase in occupational intensity represented by the early ceramic period is typified by the SU site (Martin 1943; Martin and Rinaldo 1950; Martin and Plog 1973; Wills 1991). Five field seasons at the SU site have revealed at least 28 dwelling structures, including a possible communal house, and numerous pit features. These features extend at least 600 m along a ridge in two prominent clusters (Figure

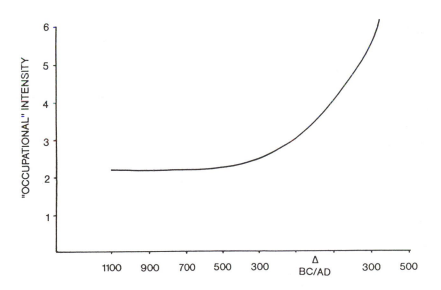

Figure 4. Occupational intensity curve for the Highland Mogollon region of west-central New Mexico (see text for details).

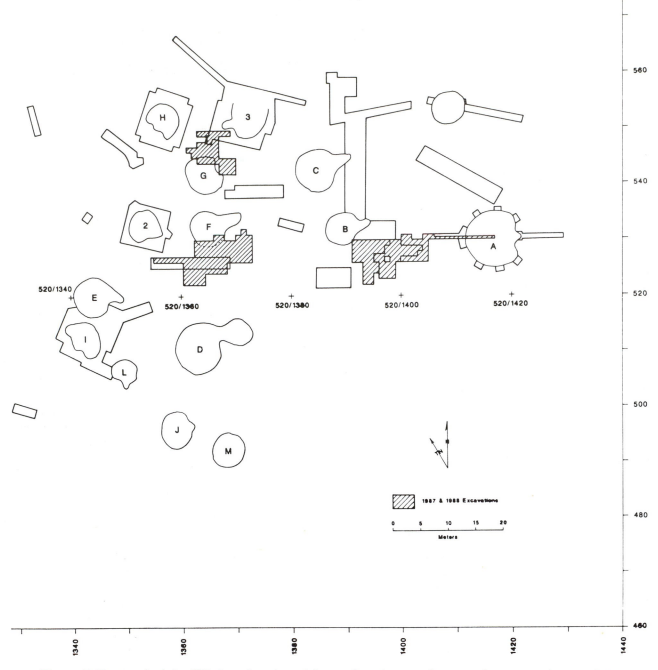

Figure 5. East end of the SU site, showing pithouse locations and excavation areas. A complete site map can be found in Martin (1943).

5). A dendrochronological cutting date of 460 A.D. was obtained from one structure and radiocarbon determinations are consistent with site occupation from about A.D. 300–500 (Wills 1989).

Features at the SU site strongly suggest a household mode of production. The pithouses are quite large, averaging nearly 40 m² in floor area and ranging to almost 80 m². These structures are also deep, some-

times exceeding a meter, and had massive post-support systems for the roof. In both size and labor-expenditure, these dwellings indicate family or extended family units. Even more striking than their size, however, is the large number and size of pit features found within these pithouses (Figure 6). The potential storage volume of these internal pits averages 2.8 m³ per house, a figure that might be extrapolated to approximately 253 kg of maize, providing enough calories for 455 person-days (see Huckell 1990:356 for estimates derived from Minnis 1985; Wetterstrom 1986; Ford 1968). Some extramural pits also exist, although many appear to have been for roasting rather than storage. Thus the average pit volume per house at the SU site could contain enough maize to provide caloric requirements for a family of five for about three months if used for that purpose.

Not all of the SU site pithouses were occupied contemporaneously, as many contained burials in the fill of abandoned houses. Fifty-four burials were recorded at the SU site, an extraordinary number for the early pithouse period in the Southwest, and 38 of these were associated with abandoned dwellings. It seems reasonable to view this large burial population as an indication of a close and long-term group affiliation with the site, that may have occurred as part of a major settlement change to year-round occupation of highland areas. The large houses and storage facilities seem to indicate extended occupation periods, potentially during winter months (see Gilman 1987). If so, then the SU site and other early pithouse settlements in the region represent a distinct contrast to the small preceramic sites occupied during the summer and fall.

A year-round residential pattern within the highland zone has interesting implications for economic production schedules. Whereas a system of seasonal mobility allows bands to take advantage of upland resource abundance in late summer and early fall, and then move to other locations in other seasons, a residential system forces people to derive a much larger portion of their annual subsistence income during a shorter period of time. In effect, the production cycles of most foraging activities (and especially those associated with annual high yield plants like pinyon pine nuts) begin to duplicate plant cultivation cycles. A decision to invest in a residential system within the uplands suggests a more temporally focused subsistence system than before, consistent with an economic strategy that is risk prone rather than risk averse.

The apparent evidence for a household mode of production and residential stability at the SU site is not found at most contemporaneous pithouse settlements in other parts of the Southwest. For example, Shabik'eschee Village in Chaco Canyon (Roberts 1929; Wills and Windes 1989) has numerous pithouses, many storage features, and a high density of artifacts, including pottery (Table 2). Unlike the SU site, however, the houses at Shabik'eschee Village are small (average floor area is 17.8 m²) and none have internal storage pits. Five of the eighteen pithouses had an "antechamber," which may have been for storage or used as an entrance, but most storage pits were outside the dwellings (Figures 7 and 8). Only one of fourteen

burials recovered from Shabik'eschee was associated with a dwelling and only one definitely belongs to the early ceramic occupation at the site. The small pithouses seem unsuited for family-sized units, food storage is apparently kept in public locations and the mortuary pattern does not suggest a strong tie to the site location. In short, data from Shabik'eschee are consistent with a communal mode of production (an extended version of this argument is found in Wills 1991).

If the expectations for conditions favoring household production are correct, then the distinctions between the SU site and Shabik'eschee Village imply a more predictable and secure resource structure for the inhabitants of the SU site. A brief comparison of environmental variation between the Mogollon Highlands and Chaco Canyon confirms this expectation. First, the mixed-conifer forest and large, permanent rivers of the Mogollon Highlands support the highest density of wild game in the Southwest today, and presumably did so in the past. The highlands also have a wide range of plant foods, including pinyon nuts, walnuts, acorns, and a variety of important aquatic plants found in bogs, cienegas, and riparian habitats (Hevly 1983). Chaco Canyon is in the San Juan Basin, a 20,000 km² area of sage grasslands with spotty distributions of pinyon-juniper woodlands. Compared to the Mogollon Highlands, the diversity and abundance of naturally available foods is much lower.

Both areas today have short growing seasons for maize cultivation but seasonal precipitation is almost double in the Mogollon Highlands (Table 3). If crops survive the threat of spring frosts, water for plant growth is much more predictable and likely in the highlands than the San Juan Basin. Modern climatic patterns indicate that in addition to a higher mean precipitation level, the highlands experience less variance around the mean (Table 3) and consequently, conditions favorable to agriculture are more evident in the highlands.

Paleoclimatic data for the early ceramic period throughout the Southwest indicate a period of higher water table levels from approximately A.D. 350 to 700

Table 2. A Comparison of Site Features between the SU site and Shabik'eschee Village.

	SU	Shabik'eschee
Estimated Number of Pithouses	35	68
Number of Excavated Pithouses	28	18
Average Floor Area (m²): per Pithouse	36.8	17.8
Average Storage Volume (m³): per Pithouse	2.8	0
Number of Burials	54	14

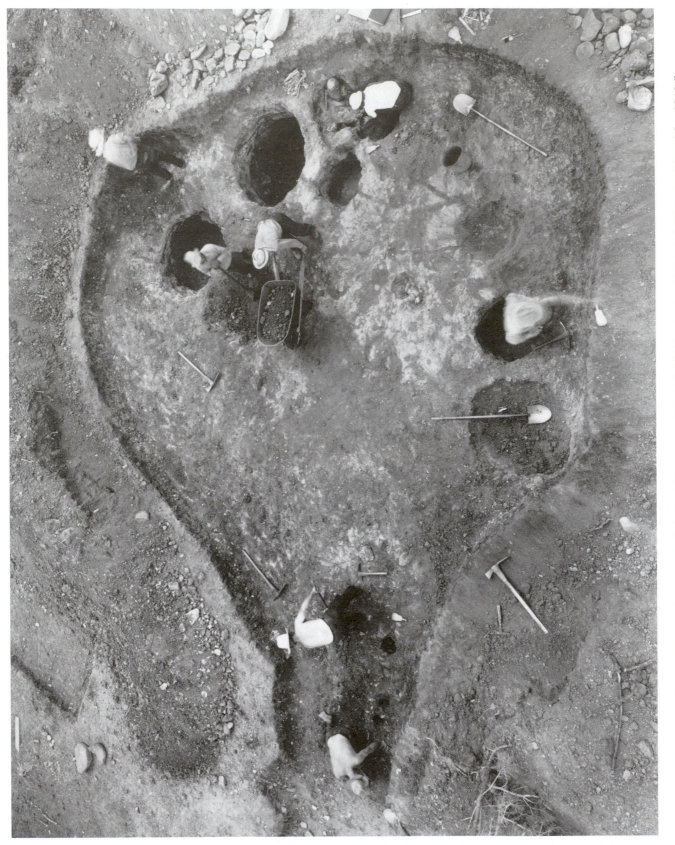

Figure 6. Excavation of Pithouse B, SU site (Courtesy of the Field Museum of Natural History, Neg. No. 88136). Notice the large interior subfloor pits.

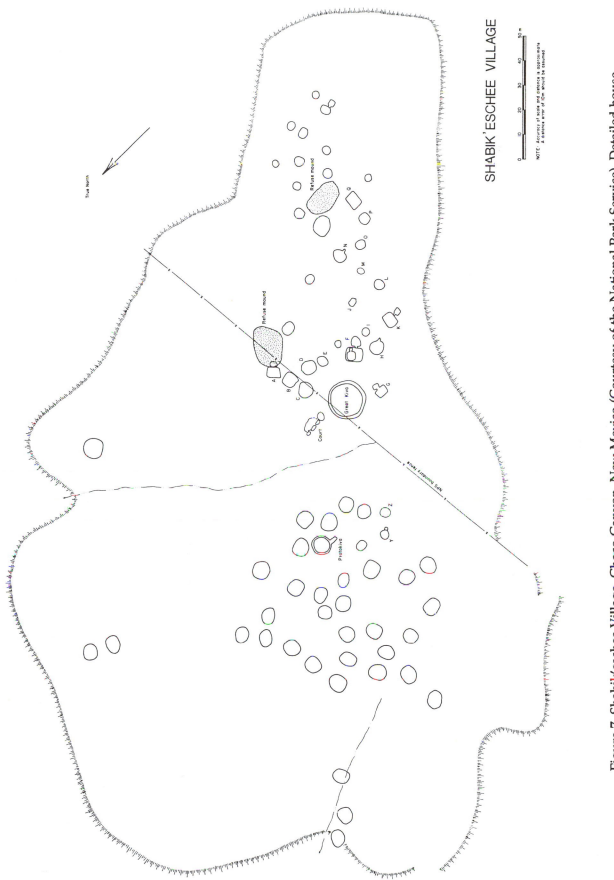

Figure 7. Shabik'eschee Village, Chaco Canyon, New Mexico (Courtesy of the National Park Service). Detailed house plans can be found in Roberts (1929).

Table 3. A Comparison of Selected Environmental Variables in the Mogollon Mountains and the San Juan Basin, New Mexico.

	Mogollon Mts.		San Juan Basin	
Average Number of Frost-free Days	118.5		155.4	
Monthly Precipation (in):	Average	SD	Average	SD
January	0.98	0.79	0.39	0.39
February	0.91	0.73	0.60	0.57
March	1.25	1.00	0.59	0.53
April	0.66	0.57	0.51	0.61
May	0.45	0.43	0.71	0.71
June	0.64	0.63	0.38	0.18
July	2.23	0.90	1.10	0.59
August	2.47	1.25	1.30	0.87
Total	9.59	5.58		

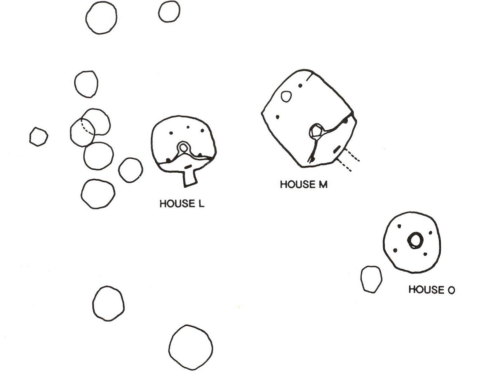

Figure 8. Detail of excavated portions of Shabik'eschee Village (see Figure 7). Notice the exterior location of probable storage features.

(Dean et al. 1985). The interval from A.D. 300 to 550 corresponds to a relatively low variability in precipitation as measured by tree-rings. This period of increasing effective moisture follows an episode between A.D. 250 and 350 when water tables dropped sharply and temporal variability increased from high levels and low variability between at least A.D. 1 and 250.

These climatic reconstructions indicate that the appearance of a household mode of production at the SU site occurred in a context of improving environmental conditions. The sudden appearance of bison in the archaeological record of the Mogollon Highlands at this time also indicates relatively high moisture regimes and extensive grasslands, as well as the availability of an extremely valuable food resource (Dick 1965; Wills 1988a).

If environmental conditions were improving, how do we account for the assumption that economic intensification indicated by household production strategies had an origin in competition for key resources? Probably because wetter conditions would increase the predictability and abundance of many wild resources, which should have encouraged their use by humans (see Earle 1980). Most important naturally available food sources in the highlands have a limited time period of availability, primarily in late summer and early fall, and this timing would not be affected by environmental change. In other words, while climatic changes may have increased the abundance of wild resources, they were still only available for a brief time. If greater abundance encouraged utilization but the time of availability remained stable, competition among human groups for resources might be an expectable development. Thus it is conceivable that adoption of risk prone economic organization took place in a context of improving economic conditions and the selective advantage in risk taking was control of a predictable and productive resource set with limited, seasonal availability.

Conclusions

The introduction of cultigens to the Southwest during the Late Archaic helped maintain economic systems with communal modes of production by allowing more effective foraging during critical periods of natural resource availability. Economic organization during the first thousand years or more of cultigen presence in the Southwest was strongly risk averse, indicated by diminishing returns and high population mobility; the adoption of plant cultivation did not lead to economies of increasing returns. A shift to risk prone economic organization during the early ceramic period was apparently based on restricted sharing and long-term investment in localities that provided access to productive and predictable natural resources. This type of shift was dictated by demographic and environmental factors that permitted some communities to take risks in production that groups in many areas could not afford. It appears that competition for access to critical resource areas came about in a context of improving climatic conditions that made intensification increasingly feasible.

Within this general developmental sequence during the Late Archaic and early ceramic periods there is good evidence for temporal and spatial variation in the economic role of plant cultivation. Pithouse settlements post-dating A.D. 200 exhibit some formal similarities over large portions of the Southwest and

are frequently treated as a developmental stage in the evolution of settled life. However, a close examination of site organization indicates that the simple presence of pithouses or pottery alone does not predict economic strategies. It seems likely that food production in the Late Archaic provided a range of useful new tactics for increasing the success of ongoing strategies. The development of restricted sharing and land tenure in the early pithouse period represented an "explosion" of economic options based on variable production tactics that increased the number of economic niches available to small groups. In upland areas, sedentism required intensive production during a short period of resource availability; while food production in desert areas could be integrated over a longer portion of the year and did not necessitate the same degree of dependency on complex scheduling decisions. Different modes of production in various parts of the Southwest likely indicate labor accomodations to concordances in natural and agricultural production cycles. Therefore, it seems that the early agricultural period in the Southwest was a time of tremendous economic flexibility and opportunity characterized by sophisticated "ecological fine-tuning" (compare with Netting 1990:40), rather than a monolithic episode characterized by such common terms as "village life" or "settling in stage."

A tendency among southwestern archaeologists to describe the early agricultural period so generally probably accounts for the absence of Southwestern data in global syntheses of agricultural origins (see Flannery 1973; Stark 1986). Nonetheless, a careful consideration of the Southwest should be useful in understanding agricultural development as a process, since foragers made an immediate commitment to the behaviors necessary to insure the survival of cultigens, and because there is a excellent chronology for this decision-making process compared to other areas.

In terms of early plant husbandry, the Southwest resembles other parts of the world in at least two ways. As in the Near East, the adoption of domesticates took place after major environmental change resulting from a period of elevated aridity that either provided new ensembles of resources and/or raised regional carrying capacities (see Bar-Yosef and Belfer-Cohen 1989; McCorriston and Hole 1991; Roberts 1991). And, like many regions, initial involvement with cultigens did not immediately lead to the emergence of economic systems predicated on agricultural surplus (see Hayden 1990; Plog 1990; Blumler and Byrne 1991).

Southwestern data may eventually contribute to our appreciation of these broad similarities through the observation that increasing reliance on cultigens, and especially the shift to household modes of production that permitted agricultural intensification, seems to have evolved in response to competition stemming from favorable conditions for both foraging and cultivation. The potential for yield does not seem to have driven changes in economic systems; but rather agricultural investment appears to have grown in response to selection for systemic economic change.

Acknowledgements

The basic argument about economic risk-taking contained herein represents an outgrowth of research that has also been reported in other places. Portions of this paper that concern the evidence for economic intensification during the Late Archaic were first developed in a presentation to a School of American Research Advanced Seminar, chaired by George Gumerman, in September of 1989. I am grateful to all the participants and Douglas Schwartz for their encouragement, and especially to Bruce Huckell, whose excellent dissertation research provided much of the data on early agriculture in southern Arizona.

Other individuals who listened patiently and gave insights freely include Patricia Crown, Jeff Dean, Steve Plog, and Bob Netting. Research at the SU site in 1988 by the University of New Mexico was sponsored by the National Geographic Society and the Wenner-Gren Foundation for Anthropological Research under a permit from the United States Forest Service.

References Cited

Adams, Karen R.
  1990 A Regional Synthesis of Zea Mays in the Prehistoric American Southwest. Paper presented at the conference entitled "Corn and Culture in the Prehistoric New World," University of Minnesota.
Adams, Jenny L.
  1989 Experimental Replication of the Use of Groundstone Tools. *The Kiva* 54(3):261-71.
Agogino, George, and Frank C. Hibben
  1958 Central New Mexico PaleoIndian Cultures. *American Antiquity* 28:422-425.
Arthur, W. Brian
  1990 Positive Feedbacks in the Economy. *Scientific American* 262:92-99.
Basehart, Harry W.
  1973 Cultivation Intensity, Settlement Patterns and Homestead Forms Among the Matengo of Tanzania. *Ethnology* 11:57-73.
Bartlett, Peggy F.
  1980 Adaptive Strategies in Peasant Agricultural Production. *Annual Review of Anthropology* 9:545-73.
Bar-Yosef, Ofer, and Anna Belfer-Cohen
  1989 The Origins of Sedentism and Farming Communities in the Levant. *Journal of World Prehistory* 3:447-498.
Baumol, William. J., and Alan S. Blinder
  1988 *Microeconomics: Principles and Policy.* Harcourt Brace Jovanovich, New York.
Beckett, Patrick H.
  1973 *Cochise Culture Sites in South Central and North Central New Mexico.* Unpublished M.A. Thesis, Department of Anthropology, Eastern New Mexico University, Portales.
Benz, Bruce F.
  1986 *Taxonomy and Evolution of Mexican Maize.* Ph.D. dissertation, University of Wisconsin. University Microfilms, Ann Arbor.
Bettinger, Robert L.
  1980 Explanatory/Predictive Models of Hunter-Gatherer Adaptation. *Advances in Archaeological Method and Theory* 3:189-255. Academic Press, New York.
Binford, Lewis R.
  1980 Willow Smoke and Dog's Tails: Hunter-gatherer Settlement Systems and Archaeological Site Formation. *American Antiquity* 45:1-17.
  1990 Mobility, Housing, and Environment: A Comparative Study. *Journal of Anthropological Research* 46:119-154.

Blumler, Mark A., and Roger Byrne
   1991 The Ecological Genetics of Domestication and the Origins of Agriculture. *Current Anthropology* 32:23-41.
Cancian, Frank
   1979 *The Innovator's Situation: Upper-Middle-Class Conservatism in Agricultural Communities.* Stanford University Press, Palo Alto.
Carmichael, David L.
   1986 *Archaeological Survey in the Southern Tularosa Basin of New Mexico.* Historic and Natural Resources Report No. 3. Environmental Management Office, United States Army, Fort Bliss, Texas.
Cordell, Linda S.
   1984 *Prehistory of the Southwest.* Academic Press, New York.
Cordell, Linda S., and George J. Gumerman.
   1989 *Dynamics of Southwest Prehistory.* Smithsonian Institution Press, Washington, D.C.
Crown, Patricia L.
   1990 The Hohokam of the American Southwest. *Journal of World Prehistory* 4:223-255.
Cutler, Hugh C., and Thomas W. Whitaker
   1961 History and Distribution of the Cultivated Cucurbits in the Americas. *American Antiquity* 26:469-485.
Dean, Jeffrey S., Robert C. Euler, George J. Gumerman, Fred Plog, Richard H. Hevly, and Thor N.V. Karlstrom
   1985 Human Behavior, Demography, and Paleoenvironment on the Colorado Plateaus. *American Antiquity* 50:537-554.
Decker, D. S.
   1984 Origin(s), Evolution, and Systematics of *Cucurbita pepo* (Cucurbitaceae). *Economic Botany* 42:4-15.
Dick, Herbert W.
   1965 *Bat Cave.* School of American Research Monograph No. 27, Santa Fe.
Donaldson, Marica L.
   1984 Botanical Remains from Sheep Camp and Ashislepah Shelters. In *Archaic Prehistory and Paleoenvironments in the San Juan Basin, New Mexico: The Chaco Shelters Project,* edited by A.H. Simmons, pp. 167-185. University of Kansas Museum of Anthropology, Project Report Series No. 53. Lawrence, Kansas.
Earle, Timothy K.
   1980 A Model of Subsistence Change. In *Modeling Change in Prehistoric Subsistence Economies,* edited by T.K. Earle and A.L. Christenson, pp. 1-29. Academic Press, New York.
Eddy, Frank W., and M. E. Cooley
   1983 *Cultural and Environmental History of Cienega Valley, Southeastern Arizona.* Anthropological Papers of the University of Arizona No. 43. Tucson.
Flannery, Kent V.
   1972 The Origins of the Village as a Settlement Type in Mesoamerica and the Near East: A Comparative Study. In *Man, Settlement and Urbanism,* edited by P. Ucko, R. Tringham, and G.W. Dimbleby, pp. 321-336. Gerald Duckworth & Co., London.
   1973 The Origins of Agriculture. *Annual Review of Anthropology* 2:271-309.
   1986 *Guilá Naquitz: Archaic Foraging and Early Agriculture in Oaxaca, Mexico.* Academic Press, New York.
Fields, R. C., and J. S. Girard
   1983 *Investigations at Site 32 (41EP325), Keystone Dam Project.* Reports of Investigations No. 21. Prewitt and Associates, Inc., Austin, Texas.
Fish, Suzanne K., Paul R. Fish, and John Madsen
   1990 Sedentism and Settlement Mobility in the Tucson Basin Prior to A.D. 1000. In *Perspectives on Southwestern Prehistory,* edited by P.E. Minnis and C.L. Redman, pp. 76-91. Westview Press, Boulder, Colorado.

Ford, Richard I.

1968 *An Ecological Analysis Involving the Population of San Juan Pueblo, New Mexico*. Ph.D. dissertation, Department of Anthropology, University of Michigan. University Microfilms, Ann Arbor.

1981 Gardening and farming before AD 1000: Patterns of Prehistoric Cultivation North of Mexico. *Journal of Ethnobiology* 1:6-27.

1984 Ecological Consequences of Early Agriculture in the Southwest. In *Papers on the Archaeology of Black Mesa, Arizona, II*, edited by S. Plog and S. Powell, pp. 127-38. Southern Illinois Press, Carbondale.

Galinat, Walton C.

1985 Domestication and Diffusion of Maize. In *Prehistoric Food Production in North America*, edited by R.I. Ford, pp. 245-78. Anthropology Papers No. 75, Museum of Anthropology, University of Michigan, Ann Arbor.

1988 The Origin of Maiz de Ocho. *American Anthropologist* 90:682.

Gilman, Patricia A.

1987 Architecture as Artifact: Pit Structures and Pueblos in the American Southwest. *American Antiquity* 52:538-564.

Green, Stanton W.

1980 Toward a General Model of Agricultural Systems. *Advances in Archaeological Method and Theory* 3:311-355.

Hall, Steven A.

1985 Quaternary Pollen Analysis and Vegetational History of the Southwest. In *Pollen Records of the Late Quarternary North American Sediments*, edited by V.M. Bryant and R.G. Holloway, pp. 95-123. American Association of Stratigraphic Palynologists Foundation, Dallas.

Hard, Robert J.

1990 Agricultural Dependence in the Mountain Mogollon. In *Perspectives on Southwestern Prehistory*, edited by Paul Minnis and Charles L. Redman, pp. 135-149. Westview Press, Boulder, Colorado.

Haury, Emil W.

1950 *The Archaeology and Stratigraphy of Ventana Cave*. University of Arizona and New Mexico Presses, Tucson and Albuquerque.

Hawkes, Kristen

1990 Why do Men Hunt? Benefits for Risky Choices. In *Risk and Uncertainty in Tribal and Peasant Economies*, edited by E. Cashdan, pp. 145-166. Westview Press, Boulder, Colorado.

Hawkes, K., H. Kaplan, K. Hill, and A.M. Hurtado

1987 Ache at the Settlement: Contrasts between Farming and Foraging. *Human Ecology* 15:133-161.

Hayden, Brian

1981 Research and Development in the Stone Age: Technological Transitions Among Hunter-Gatherers. *Current Anthropology* 22:519-48.

Hayden, Brian

1990 Nimrods, Piscators, Pluckers and Planters: The Emergence of Food Production. *Journal of Anthropological Archaeology* 9:31-69.

Hegmon, Michelle

1989 Risk Reduction and Variation in Agricultural Economies: A Computer Simulation of Hopi Agriculture. *Research in Economic Anthropology* 11:89-121.

Hevly, R. H.

1983 High-altitude Biotic Resources, Paleoenvironments, and Demographic Patterns: Southern Colorado Plateaus, A.D. 500 - 1400. In *High Altitude Adaptations in the Southwest*, edited by J. Winter, pp. 22-40. Cultural Resources Management Report No. 2, U.S. Forest Service, Albuquerque.

Huckell, Bruce B.

1984 *The Archaic Occupation of the Rosemont Area, Northern Santa Rita Mountains, Southeastern Arizona*. Cultural Resource Management Division, Arizona State Museum, Archaeological Series No. 147, Vol. I, Tucson.

Huckell, Bruce B.
  1988 Late Archaic Archaeology of the Tucson Basin: A Status Report. In *Recent Research on Tucson Basin Prehistory: Proceedings of the Second Tucson Basin Conference,* edited by William Doelle and Paul Fish, pp. 57-80. Institute For American Research Papers No. 10, Tucson.

Huckell, Bruce B.
  1990 *Late Preceramic Farmers-foragers in Southeastern Arizona: A Cultural and Ecological Consideration of the Spread of Agriculture into the Arid Southwestern United States.* Unpublished Ph.D. dissertation, Department of Arid Lands Resource Sciences, University of Arizona, Tucson.

James, Steven R.
  1990 Monitoring Archaeofaunal Changes During the Transition to Agriculture in the American Southwest. *The Kiva* 56:25-44.

Kaplan, Hillard, and Kim Hill
  1985 Food Sharing among Ache Foragers: Tests of Explanatory Hypotheses. *Current Anthropology* 26:223-237.

Kaplan, Lawrence
  1981 What is the Origin of the Common Bean? *Economic Botany* 35:240-254.

Katzenberg, M. Anne, and Jane H. Kelley.
  1988 Stable Isotope Analysis of Prehistoric Bone from the Sierra Blanca Region of New Mexico. Paper presented at the 1988 Mogollon Conference.

King, Frances B.
  1987 *Prehistoric Maize in Eastern North America: An Evolutionary Perspective.* Ph.D. dissertation, Department of Agronomy, University of Illinois. University Microfilms, Ann Arbor.

Lancaster, James W.
  1983 *An Analysis of Manos and Metates from the Mimbres Valley, New Mexico.* Unpublished M.A. thesis, Department of Anthropology, University of New Mexico, Albuquerque.

Lang, Richard W.
  1980 *Archaeological Investigations at Pueblo Agricultural Site, and Archaic and Puebloan Encampments on the Rio Ojo Caliente, Rio Arriba County, New Mexico.* School of American Research, Santa Fe.

LeBlanc, Steven A.
  1982 The Advent of Pottery in the Southwest. In *Southwestern Ceramics: A Comparative Review,* edited by A. Schroeder, pp. 107-128. The Arizona Archaeologist No. 15, Phoenix.

Lee, Richard B.
  1990 Primitive Communism and the Origin of Social Inequality. In *The Evolution of Political Systems: Sociopolitics in Small-scale Societies,* edited by Steadman Upham, pp. 225-246. Cambridge University Press.

Leonard, Robert D.
  1989 Resource Specialization, Population Growth, and Agricultural Production in the American Southwest. *American Antiquity* 54:491-503.

Lightfoot, Kent, and Gary Feinman
  1982 Social Differentiation and Leadership in Early Pithouse Villages in the Mogollon Region of the American Southwest. *American Antiquity* 47:64-86.

Long, A., R.I. Ford, D.J. Donahue, A.T. Jull, T.W. Linick, and T. Zabel
  1986 Tandem Accelerator Dating of Archaeological Cultigens. Paper presented at the 51st Annual Meeting of the Society for American Archaeology, New Orleans.

Mabry, Jonathan B.
  1990 *A Late Archaic Occupation at AZ AA:12:105 (ASM).* Center for Desert Archaeology, Technical Report No. 90-6, Tucson.

MacNeish, Richard S.
  1989 Defining the Archaic Chihuahua Tradition. *Annual Report of the Andover Foundation for Archaeological Research.* Andover, Massachusetts.

Martin, Paul S.
  1943 *The SU Site. Excavations at a Mogollon Village, Western New Mexico.* Fieldiana: Anthropology, Vol. 32, No. 1.

Martin, Paul S., and John B. Rinaldo

1950 *Sites of the Reserve Phase, Pine Lawn Valley, Western New Mexico.* Fieldiana: Anthropology, Vol. 38, No. 3.

Martin, Paul S., and Fred Plog

1973 *The Archaeology of Arizona.* Natural History Press, New York.

Matson, R.G., and Brian Chisholm

1986 Basketmaker II Subsistence: Carbon Isotopes and Other Dietary Indicators from Cedar Mesa, Utah. Paper presented at the Third Anasazi Conference.

McCorriston, Joy, and Frank Hole

1991 The Ecology of Seasonal Stress and the Origins of Agriculture in the Near East. *American Anthropologist* 93:46-69.

Minnis, Paul E.

1985 Domesticating People and Plants in the Greater Southwest. In *Prehistoric Food Production in North America,* edited by R.I. Ford, pp. 309-340. Museum of Anthropology, University of Michigan, Papers No. 75, Ann Arbor.

Nelson, Margaret C.

1990 Comments: Sedentism, Mobility, and Regional Assemblages: Problems Posed in the Analysis of Southwestern Prehistory. In *Perspectives on Southwestern Prehistory,* edited by P.E. Minnis and C.L. Redman, pp. 150-156. Westview Press, Boulder.

Netting, Robert McC.

1990 Population, Permanent Agriculture, and Polities: Unpacking the Evolutionary Portmanteau. In *The Evolution of Political Systems: Sociopolitics in Small-Scale Sedentary Societies,* edited by Steadman Upham, pp. 21-61. Cambridge University Press.

O'Laughlin, Thomas C.

1980 *The Keystone Dam Site and Other Archaic and Formative Sites in Northwest El Paso, Texas.* Publications in Anthropology No. 8, El Paso Centennial Museum.

O'Leary, B.L., and Jan V. Biella

1987 Archaeological Investigations at the Taylor Ranch Site. In *Secrets of a City: Papers on Albuquerque Area Archaeology,* edited by A.V. Poore and J. Montgomery, pp. 192-208. The Archaeological Society of New Mexico: 13, Albuquerque.

Plog, Stephen

1990 Agriculture, Sedentism, and Environment in the Evolution of Political Systems. In *The Evolution of Political Systems: Sociopolitics in Small-scale Sedentary Societies,* edited by Steadman Upham, pp. 177-202. Cambridge University Press.

Powell, Shirley

1990 Sedentism or Mobility: What do the Data Say? What Did the Anasazi Do? In *Perspectives on Southwestern Prehistory,* edited by P.E. Minnis and C.L. Redman, pp. 92-102. Westview Press, Boulder.

Price, T.D., and J.A. Brown (editors)

1985 *Prehistoric Hunter-Gatherers: The Emergence of Cultural Complexity.* Academic Press, New York.

Redding, Richard W.

1986 A General Explanation of Subsistence Change: From Hunting and Gathering to Food Production. *Journal of Anthropological Archaeology* 7:56-97.

Reinhard, K.J., J.R. Ambler, and M. McCuffie

1985 Diet and Parasitism at Dust Devil Cave. *American Antiquity* 50:819-824.

Roberts, Frank H.H.

1929 *Shabik'eschee Village: A Late Basketmaker Site in the Chaco Canyon, New Mexico.* Bulletin 92. Bureau of American Ethnology, Smithsonian Institution, Washington, D.C.

Roberts, Neil

1991 Late Quaternary Geomorphological Change and the Origins of Agriculture in South Central Turkey. *Geoarchaeology* 6:1-26.

Roth, Barbara J.

1989 *Late Archaic Settlement and Subsistence in the Tucson Basin.* Unpublished Ph.D. dissertation, Department of Anthropology, University of Arizona, Tucson.

Sahlins, Marshall
  1972 *Stone Age Economics*. Aldine, Chicago.
Sayles, E.B.
  1983 *The Cochise Cultural Sequence in Southeastern Arizona*. Anthropological Papers of the University of Arizona No. 42, Tucson.
Schlanger, Sarah H.
  1990 Artifact Assemblage Composition and Site Occupation Duration. In *Perspectives on Southwestern Prehistory*, edited by P.E. Minnis and C. L. Redman, pp. 103-121. Westview Press, Boulder.
Schiffer, Michael
  1987 *Formation Processes of the Archaeological Record*. University of New Mexico Press, Albuquerque.
Schoeninger, Margaret J.
  1990 Carbon Isotope Evidence for Diet. Paper presented at the 2nd Annual Southwest Symposium, Albuquerque.
Seaman, Timothy J.
  1988 Phase II Analysis Results. In *The Border Star Survey: Toward an Archaeology of Landscapes*, edited by T.J. Seaman, W.H. Doleman, and R.C. Chapman, pp. 121-136. Office of Contract Archaeology, University of New Mexico, Albuquerque.
Shackley, M. Steven
  1985 Lithic Raw Material Procurement and Hunter-Gatherer Mobility in the Southwest Archaic. Paper presented at the 50th Annual Meeting of the Society for American Archaeology, Denver.
Simmons, Allan H.
  1986 New Evidence for the Early Use of Cultigens in the American Southwest. *American Antiquity* 51:73-88.
Smiley, F.E.
  1985 *The Chronometrics of Early Agricultural Sites in the American Southwest*. Ph.D. dissertation, University of Michigan. University Microfilms, Ann Arbor.
Smiley, F.E., and William J. Parry
  1990 Early, Intensive, and Rapid: Rethinking the Agricultural Transition in the Northern Southwest. Paper presented at the 55th Annual Meeting of the Society for American Archaeology, Las Vegas.
Smith, Eric A., and Robert Boyd
  1990 Risk and reciprocity: Hunter-gatherer sociecology and the problem of collective action. In *Risk and Uncertainty in Tribal and Peasant Economies*, edited by E. Cashdan, pp. 167-192. Westview Press, Boulder.
Stark, Barbara
  1986 Origins of Food Production in the New World. In *American Archaeology Past and Future*, edited by D. Meltzer, D. Fowler, and J. Sabloff, pp. 277-332. Smithsonian Institution Press, Washington, D.C.
Toll, Mollie S., and Anne C. Cully
  1984 Archaic Subsistence and Seasonal Round. In *Economy and Interaction Along the Lower Chaco River*, edited by P. Hogan and J. Winter, pp. 385-392. Office of Contract Archaeology, University of New Mexico, Albuquerque.
Upham, Steadman, Richard S. MacNeish, W.C. Galinat, and C.M Stevenson
  1987 Evidence Concerning the Origin of Maiz de Ocho. *American Anthropologist* 89:410-419.
Van Devender, Thomas R., Julio Betancourt, and Mark Wimberly
  1984 Biogeographic Implications of a Packrat Midden Sequence from the Sacramento Mountains, South-Central New Mexico. *Quarternary Research* 22:344-360.
Vivian, R. Gwinn
  1974 Conservation and Diversion: Water-control Systems in the Anasazi Southwest. In *Irrigation's Impact on Society*, edited by T.E. Downing and M. Gibson, pp. 95-112. Anthropological Papers of the University of Arizona No. 25. Tucson.
Waters, Michael R.
  1986 *The Geoarchaeology of Whitewater Draw, Arizona*. Anthropological Papers of the University of Arizona No. 45. Tucson.

Wetterstrom, Wilma

    1986 *Food, Diet, and Population at Prehistoric Arroyo Hondo Pueblo, New Mexico.* School of American Research Press, Santa Fe.

Wilk, Richard R., and Robert McC. Netting

    1984 Households: Changing Forms and Functions. In *Households: Comparative and Historical Studies of the Domestic Group,* edited by R. Netting, R. Wilk, and E. Arnould, pp. 1-28. University of California Press, Berkeley.

Wills, W.H.

    1988a *Early Agriculture in the American Southwest.* School of American Research Press. Santa Fe.

    1988b Early Agriculture and Sedentism in the American Southwest: Evidence and Interpretations. *Journal of World Prehistory* 2:445-488.

    1991 Organizational Strategies and the Emergence of Prehistoric Villages in the American Southwest. In *Between Bands and States,* edited by Susan A. Gregg, pp. 161-180. Occasional Paper No. 9, Center for Archaeological Investigations, Southern Illinois University at Carbondale.

Wills, W.H., and Thomas C. Windes

    1989 Evidence for Population Aggregation and Dispersal during the Basketmaker III Period in Chaco Canyon, New Mexico. *American Antiquity* 54:347-369.

Winterhalder, Bruce

    1981 Optimal Foraging Strategies and Hunter-Gatherer Research in Anthropology: Theory and Models. In *Hunter-Gatherer Foraging Strategies: Ethnographic and Archaeological Analyses,* edited by B. Winterhalder and E. Smith, pp. 13-35. University of Chicago Press.

    1990 Open Field, Common Pot: Harvest Variability and Risk Avoidance in Agricultural and Foraging Societies. In *Risk and Uncertainty in Tribal and Peasant Economies,* edited by E. Cashdan, pp. 67-88. Westview Press, Boulder.

Woosely, Anne I.

    1980 Agricultural Diversity in the Prehistoric Southwest. *The Kiva* 45:317-336.

Index